# Lecture Notes in Bioinformatics 7699

Subseries of Lecture Notes in Computer Science

More information about this series at http://www.springer.com/series/5381

Oded Maler · Ádám Halász
Thao Dang · Carla Piazza (Eds.)

# Hybrid
# Systems Biology

Second International Workshop, HSB 2013
Taormina, Italy, September 2, 2013 and
Third International Workshop, HSB 2014
Vienna, Austria, July 23–24, 2014
Revised Selected Papers

 Springer

*Editors*
Oded Maler
CNRS-VERIMAG
University of Grenoble-Alpes (UGA)
Grenoble
France

Ádám Halász
West Virginia University
Morgantown, WV
USA

Thao Dang
CNRS-VERIMAG
University of Grenoble-Alpes (UGA)
Grenoble
France

Carla Piazza
University of Udine
Udine
Italy

ISSN 0302-9743          ISSN 1611-3349   (electronic)
Lecture Notes in Bioinformatics
ISBN 978-3-319-27655-7          ISBN 978-3-319-27656-4   (eBook)
DOI 10.1007/978-3-319-27656-4

Library of Congress Control Number: 2015958329

LNCS Sublibrary: SL8 – Bioinformatics

Printed on acid-free paper

This Springer imprint is published by SpringerNature
The registered company is Springer International Publishing AG Switzerland

# Preface

This volume contains expanded versions of research papers and tutorials presented at the Hybrid Systems Biology (HSB) workshops that took place in Taormina, Italy, in 2013 and Vienna, Austria, in 2014. To put these works and workshops in context, let us reflect a bit about systems biology, which is a term overloaded with meanings.

Recent years have seen a tremendous increase in the capability to conduct high-throughput experiments in the life sciences, leading to approaches nowadays summarized as "systems biology." However, the promise to mechanistically understand complex relationships as present, for example, in multi-factorial diseases has not been realized and the medical benefits seem to be meager compared with the cost of experimentation and the volume of scientific publications. In our opinion, this is at least partly due to a lack of a comparable progress in the conceptual system-level modeling domain.

The area of systems biology draws into the life sciences researchers from many other disciplines (mathematics, physics, engineering, and computer science), who are often more fluent in certain types of abstract modeling and reasoning than the average biologist. It is generally hoped that such an interdisciplinary collaboration will increase the convergence to useful and clinically relevant models, will help reduce the cost of experimentation (which is considered as the main limiting factor in biological research), and facilitate the transfer of research results toward clinical applications. However, there are bottlenecks in this ideal flow chart that hinder rapid progress. One of them is the interaction between the modeling researcher (the modeler henceforth) and the biologist, which often falls into one of two extremes, depending on who dominates the collaboration:

1. When it is the biologist, the modeler helps him or her in solving one particular problem (which is good by itself) but the outcome of the process has no significant generality in terms of methodology and computer-aided tools, and a similar work should be done almost from scratch for the next problem. Moreover, the modeler will tend to accept the set of abstractions and observational resolution of the biologist, thus inheriting the communal bias of the latter, which is often accidental, reducing the chance of real new theoretical insights.
2. The other extreme is when the modeler is more or less independent and uses biology as yet another case-study for his or her favorite formalisms and techniques. These will be applied to biological problems, not always questioning their adequacy, and sometime giving priority to the evaluation standards of the modeler's technical community over biological significance.

As a result we see numerous conferences in the style of "X and Systems Biology," whose impact on the practice of biological research is rather limited. This is not a criticism of any individual researcher – we are all influenced by the boundary conditions set by the structure of academic scientific disciplines. And, of course, not all interactions are like this and there are successful applications of mathematics and information technology that have had an impact on the practice of biological research.

Ideally one would like to bring the contribution of abstract modeling and pragmatic mathematical analysis to the *core* of the biological scientific activity and embed it more tightly in the hypothesis–experiment loop. It is important to note that in our modern times, the mathematical support that we would like to provide biologists with is associated with software tools. Such tools implement the mathematical know-how in software similar to the computer-aided design (CAD) and simulation tools that make complex engineered systems possible, from cars and airplanes to chips and new materials.

To avoid a misunderstanding, let us first specify what is outside the scope of our intentions. We are certainly not talking about application of computer science to genomics and gene sequencing. We are not primarily interested in high-throughput experiments, big data, and machine learning, although such techniques will eventually have their (modest) place in the process of model building. There are more things that come to mind as candidates for exclusion but we need not be exhaustive.

The research directions that we want to encourage are concerned with building dynamic models of biological phenomena, from the cell level and above, and analyzing them using a variety of computational methods to debug and explore models in silico as much as possible, avoiding useless and costly experiments.

Some of these things have been done for ages in different areas and communities, for example in the theory of chemical reactions or in the study of population dynamics, and thus it is important to stress why now is a good time to re-initialize and regroup these activities, partly based on developments in hybrid systems research.

The notion of a dynamical system has evolved greatly in the second half of the twentieth century, although the specific term "dynamical system" has not been used explicitly in these developments. The computer and the brain gave rise to models based on discrete (logical, qualitative) state variables and quantity-free transition dynamics, that is, automata that often operate on a logical (not metric) time scale. Such models underlie almost everything in computers, and are used, for example, to design complex digital circuits consisting of zillions of transistors. Naturally, already in the early days, qualitative models of genetic regulatory networks based on networks of Boolean automata were proposed.

The last decades saw a kind of confluence between the classic continuous dynamical systems based on differential equations and discrete event systems in the framework of hybrid systems. Starting initially as a meeting point between computer scientists working on the verification of systems such as communication protocols or digital hardware, and control engineers interested in the design of systems such as airplanes, robots, or industrial plants, the domain gave rise to a relatively unified view of hybrid dynamics where discrete transitions and continuous activities are interleaved. Needless to say, this style of modeling where discrete transitions are considered first-class citizens is much more efficient and intuitive for mode switching dynamics such as gene activation than the various constructs employed in continuous mathematics to express discrete changes.

In terms of model analysis, discrete and hybrid systems are not amenable in general to purely analytic techniques and hence novel ones had to be developed covering the whole spectrum from reachability-based formal verification to Monte Carlo simulation. Hybrid systems research has also led to new ways of specifying and evaluating behaviors of dynamic systems: while classic approaches focus mostly on steady-state analysis, new techniques that combine logic with quantitative measures can express more complex patterns of behaviors that occur as sequences of steps. Such methods can also explore transient behavior that are perhaps more fundamental to life than the unavoidable steady state.

It is thus believed that bringing together researchers with mathematical and computational capabilities, sharing a genuine desire to contribute to the advancement of biology, and connecting them to open-minded biologists and physicians working on problems central to the life sciences can lead to a quantum leap in the efficacy of biological research. It will hopefully lead to high-quality computer-aided methods for easily navigating in the space of hypothetical models and will drastically reduce experimentation overhead. In particular, we expect the know-how that will emerge from these activities to include:

- A better understanding of the trade-offs between different styles of modeling in terms of the complexity of analysis/simulation, faithfulness to reality, difficulty to obtain experimental data and usefulness in general.
- Improved theoretical notions concerning the formal relationships between models at different levels of abstraction and granularity: for instance, what is the relation between a continuous model and its discrete approximation, between two approximations of a spatially extended model or between stochastic and deterministic models of the same phenomenon?
- Systematic methods to abstract detailed models into simpler ones (coarse graining) or to incorporate coarse weak models inside more detailed ones. The main issue here is that coarse abstract models are less specified (under determined) and simulation and analysis methods should account for that uncertainty so as to assess the robustness of proposed models.

Wishes, visions, and ambitions are often brighter than their realizations but we hope, nevertheless, that the reader will enjoy the articles in this volume and seriously consider joining this research effort.

October 2015

Oded Maler
Ádám Halász
Thao Dang
Carla Piazza

# Organization

## Program Committee

| | |
|---|---|
| Marco Antoniotti | DISCo, Università di Milano-Bicocca, Italy |
| Ezio Bartocci | TU Wien, Austria |
| Gregory Batt | Inria Paris-Rocquencourt, France |
| Luca Bortolussi | University of Trieste, Italy |
| Thao Dang | CNRS-VERIMAG, Grenoble, France |
| Vincent Danos | CNRS, France |
| Hidde De Jong | Inria, Grenoble, France |
| Alexandre Donzé | UC Berkeley, USA |
| François Fages | Inria, Rocquencourt, France |
| Eric Fanchon | CNRS, TIMC-IMAG, France |
| Hans Geiselmann | University of Grenoble, France |
| Radu Grosu | Stony Brook University, USA |
| Ádám Halász | West Virginia University, USA |
| Thomas Henzinger | IST, Austria |
| Jane Hillston | University of Edinburgh, UK |
| Agung Julius | Rensselaer Polytechnic Institute, USA |
| Heinz Koeppl | ETH Zurich, Austria |
| Hillel Kugler | Microsoft Research, UK |
| Marta Kwiatkowska | Oxford University, UK |
| Pietro Lio | University of Cambridge, UK |
| Oded Maler | CNRS-VERIMAG, Grenoble, France |
| Bud Mishra | New York University, USA |
| Chris Myers | University of Utah, USA |
| Casian Pantea | West Virginia University, USA |
| Carla Piazza | University of Udine, Italy |
| Ricardo Sanfelice | University of Arizona, USA |
| P.S. Thiagarajan | National University of Singapore |
| Verena Wolf | Saarland University, Germany |
| David Šafránek | Masaryk University, Czech Republic |

## Additional Reviewers

François Bertaux
Milan Ceska
Frits Dannenberg
Ashutosh Gupta
Nicholas Roehner

# Contents

# Immune Response Enhancement Strategy via Hybrid Control Perspective

Hyuk-Jun Chang[1]($^{(\boxtimes)}$) and Alessandro Astolfi[2,3]

[1] School of Electrical Engineering, Kookmin University,
Seoul 136-702, Republic of Korea
hchang@kookmin.ac.kr
[2] Department of Electrical and Electronic Engineering,
Imperial College London, London SW7 2AZ, UK
a.astolfi@imperial.ac.uk
[3] DICII, Università di Roma Tor Vergata, Via del Politecnico 1, 00133 Roma, Italy

**Abstract.** We investigate a control method for disease dynamics, such as HIV and malaria, to boost the immune response using a model-based approach. In particular we apply the control method to select the appropriate immune response between Th1 and Th2 responses. The idea of state jump is introduced and discussed based on hybrid control systems. To implement the control idea we propose physically available methods for each biological system. The studies on malaria model and HIV model are supported by experimental data.

**Keywords:** Hybrid systems · State jump · Malaria · Bee venom · HIV/AIDS · Immunotherapy

## 1  Introduction

Malaria and HIV/AIDS remain lethal and prevalent infectious diseases worldwide. Malaria accounts for 300 million cases and over a million fatalities every year [38]. In 2009 the estimated number of people infected with HIV was 33.3 million and 1.8 million people died of AIDS [34].

Several recent studies have discussed model-based approaches to HIV/AIDS and malaria infection. Some examples of HIV dynamic modeling are found in [35,36] and control theoretic studies on the HIV models by the authors are reported in [5,7,33]. In [21] malaria dynamics have been modeled and model parameters have been estimated. The model has been studied in [17,22], and the immune response model has been used in different malaria models [16,26].

In this chapter we investigate a control scheme based on a model-based approach to enhance the immune response in HIV and malaria dynamics. To this end we propose the idea of using state jumps as hybrid control mechanism. We study the malaria model in [21] with the perspective of the immune boosting mechanism. We then apply this control scheme to the immune population model of [2] to select the appropriate immune response. Finally the control idea is

© Springer International Publishing Switzerland 2015
O. Maler et al. (Eds.): HSB 2013 and 2014, LNBI 7699, pp. 1–26, 2015.
DOI: 10.1007/978-3-319-27656-4_1

applied to the HIV model in [35] on the basis of the immune boosting mechanism[1] analysed by the authors in [5]. Our researches on the malaria model and the HIV model are supported by experimental data which have been published in [29] and [15].

This chapter is organised as follows. We describe the application of our control idea to the malaria model of [21] in Sect. 2, based on the research of [9]. Then we apply the developed control scheme to an immune population model in [2] with bee venom injection in Sect. 3 from our study in [8]. We recall some ideas of [5] and suggest control methods to boost the immune response in the HIV dynamic model of [35] in Sect. 4 as a summary of [10]. Finally Sect. 5 concludes this chapter by presenting future works and further remarks.

## 2   Hybrid Control Method for Malaria Dynamics

The clinical work of [29] leads us to apply the mathematical method of immune enhancement in HIV dynamics to malaria dynamics. The experimental results in [29] can be explained by the immune system analysis in this section. The study presented in this section is based on the research of the authors in [9].

### 2.1   A Summary of the Experiments in Roestenberg et al. 2009

In this subsection we briefly summary the experiments of [29].

15 healthy volunteers were with an antimalaria drug regimen, *chloroquine* when they were exposed to mosquito bites once a month for 3 months. 10 of the volunteers were assigned to a vaccine group while the rest 5 volunteers were assigned to a control group. The vaccine group was exposed to mosquitoes infected with a species of malaria parasite, *Plasmodium falciparum*. The control group was exposed to mosquitoes not infected with the malaria parasite.

After 1 month the immune response was tested by the challenge with five mosquitoes infected with the malaria parasite. Malaria parasites were not found in the blood of any of the 10 volunteers in the vaccine group even with the malaria challenge with the infected mosquitoes. Also no serious adverse events occurred in this vaccine group. However malaria developed in all 5 volunteers in the control group. Thus the inoculation of malaria parasite by infected mosquitoes can lead to protection against a malaria challenge.

### 2.2   Malaria Dynamic Model and Parameters

We consider the malaria infection model of [21] in this section. Although the immune modeling in [23] is similar to [35, 36], we use the model of [21] to show that the immune boosting idea can be applied to various types of immune system models. The model is given by

---

[1] This mechanism is also used for the treatment of chronic myeloid leukaemia (CML) in [6].

**Table 1.** Malaria model parameters and their values or ranges [21]

| Parameters | Value | Parameters | Range |
|---|---|---|---|
| $a$ | 1.28 | $x$ | 0 - 1 |
| $b$ | 1.39 | $y$ | 0 - 1 |
| $g$ | 0.04 | $c_s$ | 0.0001 - 1000 |
| $q_s$ | 0.01 | $c_n$ | 0.0001 - 1000 |
| $q_n$ | 0.6 | $s_s$ | 0.0001 - 1000 |
| $k$ | 2.3 | $s_n$ | 0.0001 - 1000 |

$$\dot{V} = aV - c_s JV - c_s xKV - c_n IV - gV, \tag{1}$$

$$\dot{F} = bF - c_s KF - c_s yJF - c_n IF - gF, \tag{2}$$

$$\dot{I} = s_n(V + F) - q_n I, \tag{3}$$

$$\dot{J} = s_s V - q_s J, \tag{4}$$

$$\dot{K} = s_s F - q_s K, \tag{5}$$

where the states $V$, $F$, $I$, $J$, and $K$ describe the populations of specific cells in a unit volume of blood and therefore are meaningful only when nonnegative.

In particular $V$ describes the concentration of P. vivax[2], $F$ the concentration of P. falciparum, $I$ the concentration of the effectors of the non-specific immune response, and $J$ and $K$ the concentrations of the effectors of the specific immune response for $V$ and $F$, respectively. The remaining parameters $a$, $b$, $g$, $q_s$, $q_n$, $x$, $y$, $c_s$, $c_n$, $s_s$, and $s_n$ are positive and constant. The values of the model parameters[3] estimated in [21] are summarised in Table 1. For a detailed explanation of the model see [21].

We consider a single-species P. falciparum case by removing the variables $V$ and $J$ of model (1-5) and taking Eqs. 2, 3 and 5 as studied in [17] because this is the clinical case studied in [29]. The term of administration of antimalarials $u$ is included by introducing a parasite killing rate $k$ and we set $k = 2.3$ consistently with published ranges of parasite reduction ratios [21]. As a result the model is given by

$$\dot{F} = (b - ku)F - c_s KF - c_n IF - gF, \tag{6}$$

$$\dot{I} = s_n F - q_n I, \tag{7}$$

$$\dot{K} = s_s F - q_s K. \tag{8}$$

If $s_n F - q_n I > 0$ (or $s_s F - q_s K > 0$), then the immunity by $I$ (or $K$) is enhanced, as discussed in Sect. 4. To achieve $s_n F - q_n I > 0$ (or $s_s F - q_s K > 0$) the state

---

[2] Human malaria is caused by four species of *Plasmodium*: P. falciparum, P. malariae, P. ovale, and P. vivax [22]. Model 1-5 includes the dynamics between P. vivax, P. falciparum and the immune system.

[3] The model parameters are based on clinical data, so the state variables in this section represent actual data.

must be moved into the set $s_n F - q_n I > 0$ (or $s_s F - q_s K > 0$) by a "state jump", which is realised by mosquito inoculation in [29].

Let $Y(t) := [F(t), I(t), K(t)]^T$ and let model (6-8) be

$$\dot{Y} = G(Y, u). \tag{9}$$

In this section we consider infective mosquito bites to realise the experiment of [29] *in silico*. One mosquito bite corresponds to a total of 300,000 malaria parasites (primary merozoites) [16]. The experiment in [29] uses 36 - 45 infected mosquito bites so we assume 40 bites in each exposure to subject. Note that the state variables used in [21] represent per-microliter densities and we assume that the total human blood volume is 4,500(ml). Thus we set $Y^* = [2.67, 0, 0]^T$. Also we consider the initial state $Y(0) = [0, 0, 0]^T$, which represents the status of the subjects of the experiment in [29]. We set the drug treatment period $T_d = 90$ and the state jump treatment period $T_t = 63$ consistently with the experiment in [29]. As in [21] we use $c_s = 0.1$, $c_n = 0.1$, $s_s = 0.1$, and $s_n = 0.1$.

## 2.3   Open-Loop Control

In this subsection we consider open-loop control for the model 9 and the exposure of infective mosquito bites is carried out at $t = 7, 35, 63$, which corresponds to the case of [29]. Thus the main purpose of the open-loop control of this section is to show the effect of the experiment of [29] on the model.

The suggested control is given by

$$\dot{Y}(t) = G(Y(t), 1), \qquad \text{if } t \in [0, T_d] \text{ and } t \notin S_t,$$
$$\dot{Y}(t) = G(Y(t), 0), \qquad \text{if } t > T_d,$$
$$Y^+ = Y + Y^*, \qquad \text{if } t \in S_t,$$

where $S_t = \{7, 35, 63\}$.

Figure 1 shows the results of the application of the proposed open-loop control procedure. The specific immune effector $K$ is boosted up to 0.602 at $t = 66.75$(day) while the non-specific immune effector $I$ is not. This is because the difference in the decay rates, $q_s$ and $q_n$. Thus immune effector of lower decay rate such as memory B cells [21] are more enhanced by the given control method.

The resulting $(I, F)$ and $(K, F)$ trajectories are displayed in Fig. 2. This shows that each trajectory enters into the immune increasing area (i.e. $s_n F - q_n I > 0$ and $s_s F - q_s K > 0$) by the action of state jumps. In the simulation without state jumps, the state variables $F$, $I$, and $K$ are zeros. Accordingly the state at $t = 100$(day) of the simulation is $[0, 0, 0]^T$ which implies that the immune response is not enhanced, consistently with the result of [29].

In order to verify the robustness of the open-loop control we perform the computer simulations for unknown control parameters, $Y^*$ and $S_t$. Figure 3 shows the simulation results of the control strategy applied to 40 cases with different parameters which are generated by random perturbations from the nominal values of $Y^*$ and $S_t$. $Y^*(1)$ is perturbed up to $\pm 50\%$ and each inoculation timing

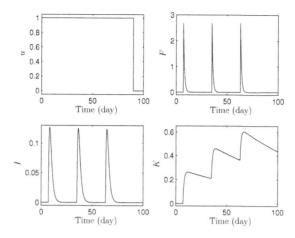

**Fig. 1.** Results of the application of the open-loop control strategy to model (9).

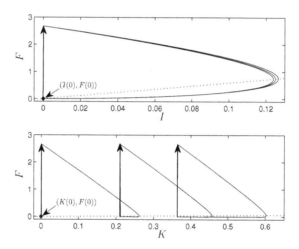

**Fig. 2.** The $(I, F)$ and $(K, F)$ trajectories resulting from the open-loop control strategy. The solid lines in each graph indicate the $(I, F)$ and $(K, F)$ trajectory, respectively. The dotted lines in each graph indicate the sets $s_n F - q_n I = 0$ and $s_s F - q_s K = 0$, respectively.

in $S_t$ is perturbed $\pm 3.5$(days). In spite of diverse control parameters the average of the maximally boosted $K$ during the control procedure is 0.5905 with the standard deviation 0.0904. This indicates that the suggested open-loop control has significantly large robust margin against control parameter uncertainty.

*Remark 1.* When the $(I, F)$ trajectory (or $(K, F)$ trajectory) passes through the set $s_n F - q_n I = 0$ (or $s_s F - q_s K = 0$), $dI/dF = 0$ (or $dK/dF = 0$). This implies that if we obtain a $(I, F)$ trajectory (or $(K, F)$ trajectory) with a point in which

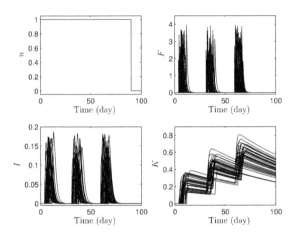

**Fig. 3.** Results of the application of the open-loop control strategy to model (9) for 40 different control parameters conditions.

$dI/dF = 0$ (or $dK/dF = 0$) from experimental data, then we could estimate the ratio $s_n/q_n$ (or $s_s/q_s$).

We now state a proposition from [9] related to the boosting level achievable by the hybrid control technique (see [9] for the proof).

**Proposition 1.** Consider the model (9) with $Y^* = [F_M, 0, 0]^T$ for some positive $F_M$ and assume $b - k - g \le r_m < 0$ and $u = 1$. Also assume that the state jump is carried out if $0 \le F \le F_m$ for some positive $F_m$. Then $F_m s_s/q_s < K(t) < (F_M + F_m)s_s/q_s$ and $F_m s_n/q_n < I(t) < (F_M + F_m)s_n/q_n$ for $t > t_F$ for some positive $t_F$.

### 2.4  Closed-Loop Control

In this subsection we construct a feedback control strategy to maintain $s_s F - q_s K \ge 0$ for $0 < t \le T_t$ so that we can enhance the immune effector $K$ which has lower decay rate than $I$. To implement this control strategy, the corresponding hybrid system is given by

$$\dot{Y}(t) = F(Y(t), 1), \quad \text{if } t \in [0, T_d],$$
$$\dot{Y}(t) = F(Y(t), 0), \quad \text{otherwise,}$$
$$Y^+ = Y + Y^*, \quad \text{if } s_s F - q_s K < 0 \text{ and } t \in [0, T_t].$$

The result of the control action is shown in Figs. 4 and 5. The total number of mosquito inoculations is 24 in this closed-loop control, while it is 3 in the open-loop control in Subsect. 2.3. However the boosting level of $K$ by closed-loop control is much higher than that of $K$ by open-loop control. (Compare the graph of $K$ in Fig. 4 with that of $K$ in Fig. 1.) Fig. 5 shows the $(I, F)$ and $(K, F)$ trajectories. In the lower graph in Fig. 5 the $(K, F)$ trajectory stays within the set $s_s F - q_s K \ge 0$ for $0 < t \le T_t$.

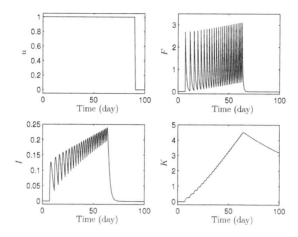

**Fig. 4.** Results of the application of the closed-loop control strategy to model (9).

## 3 Hybrid Control Method for Bee Venom Immunotherapy

The hybrid control method with state jump can be applied to dynamic model of bee venom immunotherapy[4]. For a brief introduction of venom immunotherapy by insect sting, see [14].

The study described in this section is based on the research of the authors in [8]. We consider the immune population model of [2,3]. The model is given by

$$\dot{x}_1 = \frac{\sigma_1 x_1 p}{(1 + x_2)^2} + \frac{\pi_1 x_1}{(1 + x_2)} - \delta_1 x_1^2 - x_1, \tag{10}$$

$$\dot{x}_2 = \frac{\sigma_2 x_2 p}{(1 + x_2)} + \frac{\pi_2 x_2}{(1 + x_1 + x_2)} - \delta_2 x_1 x_2 - x_2, \tag{11}$$

$$\dot{p} = p(r - \nu_1 x_1 - \nu_2 x_2), \tag{12}$$

where the states $x_1$, $x_2$, and $p$ describe the populations of specific cells in a unit volume of blood and therefore are meaningful only when nonnegative.

In particular $x_1$ describes the concentration[5] of Th1, $x_2$ the concentration of Th2, and $p$ the concentration of pathogen. Th1-cells are involved in immune responses against intracellular pathogens such as viruses, producing cytokines such as IL-2, IL-12, and IFN-$\gamma$. Th2-cells produce mainly cytokines such as IL-3, IL-4, IL-5, and IL-13, providing defence against extracellular pathogens [27]. Approximately Th1 and Th2 cells favour the cell-mediated and the humoral immune response, respectively [3]. Note that some infectious agents have a pre-

---

[4] Besides immunotherapy, bee venom are currently used in clinical cases. Experimental results of [20] are one example.

[5] Th stands for T-helper. Th1 and Th2 are two types of T-helper cells.

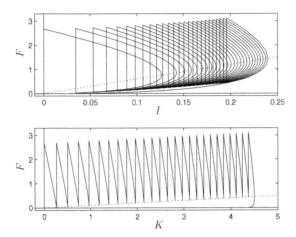

**Fig. 5.** The $(I,F)$ and $(K,F)$ trajectories resulting from the closed-loop control strategy. The solid lines in each graph indicate the $(I,F)$ and $(K,F)$ trajectories, respectively. The dotted lines in each graph indicate the sets $s_nF-q_nI = 0$ and $s_sF-q_sK = 0$, respectively.

disposition[6] to induce cell-mediated immune response while others to induce humoral immune response [2].

The parameters $\sigma_1$, $\sigma_2$, $\pi_1$, $\pi_2$, $\delta_1$, $\delta_2$, $\nu_1$, $\nu_2$, and $r$ are positive and constant. The biological meanings of the parameters are summarised in [2, Table 2] and [3, Table 1]. The typical values of the model parameters suggested in [2,3,27] are summarised in Table 2. For a detailed explanation of the model see [2,3].

**Table 2.** Venom immunotherapy model parameters and values [2,27]

| Parameters | Value | Parameters | Range |
|---|---|---|---|
| $\sigma_1$ | 2 | $\sigma_2$ | 2 |
| $\pi_1$ | 2 | $\pi_2$ | 2 |
| $\delta_1$ | 1.5 | $\delta_2$ | 0.5 |
| $\nu_1$ | 1 | $\nu_2$ | 1 |
| $r$ | 0 | | |

With these parameters, model (10-12) is a bistable system [2,3] and the two stable equilibrium points are

Point Th1: (0.6667, 0.0000, 0.0000)

Point Th2: (0.0000, 1.0000, 0.0000).

---

[6] Immune response to intracellular pathogens tends to induce Th1 dominance and resultant cellular cytolytic activity, although immune response to extracellular infection is often dominated by Th2 response, which lead to high levels of pathogen-specific immunoglobulins [2].

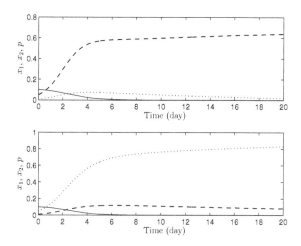

**Fig. 6.** State histories of model (10-12) from the initial state $[0.05, 0.01, 0.1]^T$ (top graph) and $[0.01, 0.05, 0.1]^T$ (bottom graph). $x_1(t)$, $x_2(t)$, and $p(t)$ are represented by dashed line, dotted line, and solid line, respectively.

To highlight this property, two trajectories of the systems are displayed in Fig. 6. The two initial points in Fig. 6 are $[0.05, 0.01, 0.1]^T$ and $[0.01, 0.05, 0.1]^T$, used in [2]. We integrate model (10-12) from these initial points for 20 days, Fig. 6 shows that each trajectory converges to one of the stable equilibrium points. Point Th1 and Point Th2 correspond to the states in which Th1 response and Th2 response dominate, respectively.

Note that comparable susceptibilities for the Fas-mediated apoptotic signal of Th1 and Th2 cells can lead to the loss of Th1-Th2 switches, crucial for the ability to select the appropriate T helper response [2]. In addition incorrect stimuli in the immune system can promote Th1 responses when Th2 responses are appropriate (or vice-versa). The major source of incorrect stimuli could be pathogen evolution to evade immune response [3]. Thus in this section the goal of hybrid control is to bring the state to within a small neighbourhood of Point Th2[7].

To this end we analyse the dynamics of model (10-12). Equations 10 and 11 can be rewritten as $\dot{x}_1 = M_1(x_1, x_2, p)x_1$ and $\dot{x}_2 = M_2(x_1, x_2, p)x_2$, respectively, where

$$M_1(x_1, x_2, p) = \frac{\sigma_1 p}{(1+x_2)^2} + \frac{\pi_1}{(1+x_2)} - \delta_1 x_1 - 1,$$

$$M_2(x_1, x_2, p) = \frac{\sigma_2 p}{(1+x_2)} + \frac{\pi_2}{(1+x_1+x_2)} - \delta_2 x_1 - 1.$$

If $M_1(x_1, x_2, p)$ and $M_2(x_1, x_2, p)$ become positive then the states $x_1$ and $x_2$ are increased by (10) and (11), respectively. Note that $M_1(x_1, x_2, p)$ and $M_2(x_1, x_2, p)$ can be increased by positive state jumps of $p$.

---

[7] The reason why the switch Th1-Th2 is guaranteed but the switch Th2-Th1 is not with the current parameters will be discussed in Sect. 5.

Figure 7 shows graphical visualisation of the equations $M_1(x_1, x_2, p) = 0$ and $M_2(x_1, x_2, p) = 0$ for the given parameters in the $(x_1, x_2, p)$ space. The light grey surface and the dark grey surface describe the set $M_1(x_1, x_2, p) = 0$ and the set $M_2(x_1, x_2, p) = 0$, respectively. Since $M_1 > 0$ above the light grey surface, the state $x_1$ increases when $(x_1, x_2, p)$ belongs to this space. In addition the state $x_2$ increases if $(x_1, x_2, p)$ stays in the space above the dark grey surface.

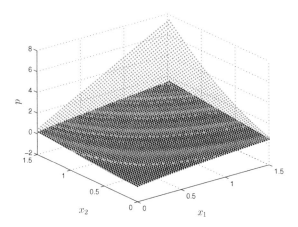

**Fig. 7.** Graphical visualisation of the state enhancing property in model (10-12). The light grey surface indicates the set $M_1 = 0$. The dark grey surface represents the set $M_2 = 0$.

To implement a control method in model (10-12) the idea of injection of bee venom in [27] is employed. In model (10-12) a injection of bee venom corresponds to a jump in the pathogen state $p$. The effect of bee venom in the dynamics can be modeled by the Dirac delta function as in [27], however we use the description of hybrid systems for the effect of state jumps.

Let $Z(t) := [x_1(t), x_2(t), p(t)]^T$ and let model (10-12) be

$$\dot{Z} = H(Z). \tag{13}$$

In the following subsections we use jump sizes of 0.1 and 1, extracted from [2] and [27], respectively.

### 3.1 Open-Loop Control

In this subsection we consider open-loop control for the model (13) and the injection of bee venom is carried out at $t = 1, 2, 3, \cdots, T_t$. The controlled system can be described by

$$\dot{Z}(t) = H(Z(t)), \qquad \text{if } t \notin S_t$$
$$Z^+ = Z + Z^*, \qquad \text{if } t \in S_t,$$

where $S_t = \{1, 2, 3, \cdots, T_t\}$ and $Z^* = (0, 0, p_j)$. In this control $T_t$ and $p_j$ correspond to the number of injections and the jump size of $p$, respectively.

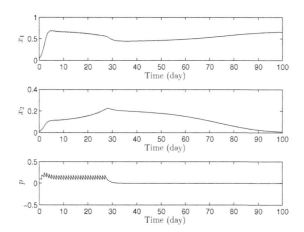

**Fig. 8.** Results of the application of the open-loop control strategy to model (13) with $S_t = \{1, 2, 3, \cdots, T_t\}$ (day). The initial state is $[0.05, 0.01, 0.1]^T$ and the jump size $p_j$ is 0.1.

Figure 8 shows the results of the application of the proposed open-loop control procedure with $T_t = 27$, $p_j = 0.1$ and the initial state $[0.05, 0.01, 0.1]^T$, used in [2] and the top graph of Fig. 6. The state at $t = 100(\text{day})$ of the simulation is $[0.6561, 0.0070, 0.0000]^T$. With this open-loop control the state $x_2$ is not enhanced so that the state cannot be driven to Point Th2. In the following subsection we show that closed-loop control considering the surface $M_2 = 0$ of Fig. 7 can drive the state to Point Th2 even with the same jump size and the same number of venom injections.

### 3.2    Closed-Loop Control

In this subsection we construct a feedback control strategy to maintain $M_2(x_1, x_2, p) \geq 0$ for $0 < t \leq T_t$ so that we can enhance the immune state $x_2$. To implement this control strategy, the corresponding hybrid system is given by

$$\dot{Z}(t) = H(Z(t)), \quad \text{if } t \in [0, T_t] \text{ and } M_2(x_1, x_2, p) \geq 0,$$
$$\text{or } t > T_t,$$
$$Z^+ = Z + Z^*, \qquad \text{if } t \in [0, T_t] \text{ and } M_2(x_1, x_2, p) < 0,$$

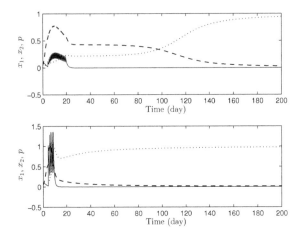

**Fig. 9.** Results of the application of the closed-loop control strategy to model (13). The initial state is $[0.05, 0.01, 0.1]^T$. $x_1(t)$, $x_2(t)$, and $p(t)$ are represented by dashed line, dotted line, and solid line, respectively. The jump sizes $p_j$ are 0.1 and 1 in the top graph and the bottom graph, respectively. The values of $T_t$ are 20 and 10 in the top graph and the bottom graph, respectively.

where $Z^* = (0, 0, p_j)$. In this subsection we study 4 control examples. Each example has its own $T_t$ and $p_j$. The result of the control action is shown in Figs. 9 and 10 where $x_1(t)$, $x_2(t)$, and $p(t)$ are represented by dashed line, dotted line, and solid line, respectively. In addition the jump sizes $p_j$ are 0.1 and 1 in the top graph and the bottom graph, respectively, in Figs. 9 and 10.

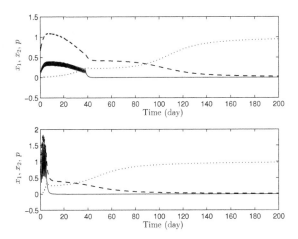

**Fig. 10.** Results of the application of the closed-loop control strategy to model (13). The initial state is $[0.6667, 0.01, 0.01]^T$ which is a point close to Point Th1. $x_1(t)$, $x_2(t)$, and $p(t)$ are represented by dashed line, dotted line, and solid line, respectively. The jump sizes $p_j$ are 0.1 and 1 in the top graph and the bottom graph, respectively. The values of $T_t$ are 40 and 5 in the top graph and the bottom graph, respectively.

In Fig. 9 the initial state is $[0.05, 0.01, 0.1]^T$, used in the open-loop control example in Subsect. 3.1. The $T_t$ values are 20 and 10 in the top graph and the bottom graph, respectively. Compared with the case of open-loop control in Fig. 8 of Subsect. 3.1, the case in the top graph shows that the closed-loop control with consideration of the surface $M_2 = 0$ is capable of leading the state into the region of attraction of Point Th2. Note that these two control examples have the same number of $p$ state jumps, namely 27.

In Fig. 10 the initial state is $[0.6667, 0.01, 0.01]^T$ which is a point sufficiently close to Point Th1. The required $T_t$ values are 40 and 5 in the top graph and the bottom graph, respectively.

# 4    Hybrid Control Method for HIV Dynamics

In this section we consider the HIV infection model of [35], which includes helper-independent CTL. The study of this section is based on the research of the authors in [10].

## 4.1    A Summary of the Experiments in Grant et al. 2010

In this subsection we briefly discuss the experiments of [15].

Antiretroviral chemoprophylaxis before exposure to HIV is expected as a promising approach for the prevention of HIV acquisition. For men and transgender women who have sex with men the current use of preexposure prophylaxis is rare, although the majority would consider such use if evidence of safety and efficacy became available.

To evaluate the clinical effect of preexposure chemoprophylaxis in men or transgender women who have sex with men, experiments have been designed and conducted in [15]. The experiments assigned 2499 HIV-negative subjects randomly to receive a combination of two oral antiretroviral drugs, i.e. emtricitabine and tenofovir disoproxil fumarate (FTC-TDF), or placebo once a day.

The subjects were followed for 3324 person-years (median, 1.2 years; maximum, 2.8 years). For the studied subjects 10 were found to have been infected with HIV at enrollment and 100 became infected during follow-up, among which 36 and 64 were in the FTC-TDF and the placebo groups respectively, implying a 44 % reduction in the incidence of HIV. Thus the experiment result concludes that oral FTC-TDF could provide protection against the acquisition of HIV infection among the subjects.

For full information of the experiment and the result data, see [15] and the Supplementary Appendix available at NEJM.org.

## 4.2    HIV Dynamic Model and Parameters

The model is given by

$$\dot{x} = \lambda - dx - \eta\beta xy, \tag{14}$$

$$\dot{y} = \eta\beta xy - ay - p_1 z_1 y - p_2 z_2 y, \tag{15}$$

$$\dot{z}_1 = c_1 z_1 y - b_1 z_1, \tag{16}$$

$$\dot{w} = c_2 x y w - c_2 q y w - b_2 w, \tag{17}$$

$$\dot{z}_2 = c_2 q y w - h z_2, \tag{18}$$

where the states $x$, $y$, $z_1$, $w$, and $z_2$ describe the populations of specific cells in a unit volume of blood and therefore are meaningful only when nonnegative.

In particular, $x$ describes the concentration of uninfected CD4 T-cells, $y$ the concentration of infected CD4 T-cells, $z_1$ the concentration of helper-independent CTLs, $w$ the concentration of CTL precursors, and $z_2$ the concentration of helper-dependent CTLs.

The quantity $\eta$, which varies between 0 and 1, describes the effect of the drug. In the presence of a control input, $\eta$ can be rewritten as $\eta = 1 - \eta^* u$, where $\eta^*$ is the maximal effect of the drug. From a control perspective the input $u$ represents the drug dose of anti-retroviral therapy. If $u = 1$ a patient receives the maximal drug therapy, while $u = 0$ means no medication. $u$ is restricted to be either 0 or 1 because the use of partially suppressive therapy, that is, $0 < u < 1$, is difficult to achieve in practice [39]. In addition, drug therapy at less than full dosage could be dangerous if carelessly applied because it may result in high probability of the emergence of drug-resistant strains of HIV [39]. The remaining parameters $\lambda$, $d$, $\beta$, $a$, $p_1$, $p_2$, $c_1$, $c_2$, $q$, $b_1$, $b_2$, and $h$ are positive and constant. For a detailed explanation of the model see [35,36].

When a patient is treated, we hope to drive its state into the region of attraction of the long-term non-progressor (LTNP). To investigate this possibility, we find the equilibrium points of the model (14-18) and obtain five equilibria [7], three of which are given in what follows.

Point A:

$$x^{(A)} = \frac{\lambda}{d}, \quad y^{(A)} = 0, \quad z_1{}^{(A)} = 0, \quad w^{(A)} = 0, \quad z_2{}^{(A)} = 0.$$

Point B:

$$x^{(B)} = \frac{\lambda c_1}{d c_1 + b_1 \eta \beta}, \qquad y^{(B)} = \frac{b_1}{c_1},$$

$$z_1{}^{(B)} = \frac{\eta \beta x^{(B)} - a}{p_1}, \qquad w^{(B)} = 0, \qquad z_2{}^{(B)} = 0.$$

Point C:

$$y^{(C)} = \frac{\theta(\eta) - \sqrt{\theta(\eta)^2 - 4\eta\beta c_2 q d b_2}}{2\eta\beta c_2 q},$$

$$x^{(C)} = \frac{\lambda}{d + \eta\beta y^{(C)}}, \quad z_1{}^{(C)} = 0, \quad w^{(C)} = \frac{h z_2{}^{(C)}}{c_2 q y^{(C)}},$$

$$z_2{}^{(C)} = \frac{y^{(C)}(c_2 \eta\beta q - c_2 a) + b_2 \eta\beta}{c_2 p_2 y^{(C)}},$$

where $\theta(\eta) = c_2(\lambda - dq) - b_2 \eta\beta$.

Under the assumption that $\eta = 1$ (i.e., no medication) the interpretation of each point is as follows (see [35] for details). Point A models the status of a person who does not have HIV. With a typical parameter set, for example from [35], this point is unstable, and this explains why it is difficult to revert a patient, once infected, back to the HIV-free status when the medication is stopped. Point B models the status of a patient for whom HIV dominates, which is stable with the same parameters. Point C models the status of an infected patient who does not progress to AIDS, which corresponds to the LTNP status. With a typical set of parameters the numbers of viral load and infected cells $y^{(C)}$ are kept low while the number of CTL precursor $w^{(C)}$ is large, which is desired. Since Point C is locally exponentially stable, with a typical set of parameters, the control goal is to drive the state near Point C.

We can regard model (14-18) as the interconnection of two subsystems: the infection dynamics and the immune system [7]. The infection dynamics are given by the two-dimensional nonlinear system described by Eqs. (14) and (15). The immune system is the three-dimensional nonlinear system described by Eqs. (16-18). The goal of the control is to enhance immunity, and this is equivalent to boosting $z_1$, $w$, and $z_2$. Particularly, helper-dependent responses (i.e. $w$ and $z_2$) must be enhanced in order to lead a HIV patient to the LTNP status (Point C), because the $w$ and $z_2$ components of Point B are zero and those of Point C are positive. $z_2$ depends upon $w$ in Eq. 18, which suggests increasing $w$ to drive the patient state into the LTNP status. Now note that Eq. 17 can be rewritten as $\dot{w} = K(x, y)w$ where

$$K(x, y) = c_2 xy - c_2 qy - b_2$$

and that $K(x, y)$ depends upon the variables $x$ and $y$ of the infection dynamics.

**Table 3.** HIV model parameters and values [35]

| Parameters | Value | Parameters | Value |
|---|---|---|---|
| $\lambda$ | 1 | $d$ | 0.1 |
| $\beta$ | 1 | $a$ | 0.2 |
| $p_1$ | 1 | $p_2$ | 1 |
| $c_1$ | 0.03 | $c_2$ | 0.06 |
| $q$ | 0.5 | $b_1$ | 0.1 |
| $b_2$ | 0.01 | $h$ | 0.1 |
| $\eta^*$ | 0.9799 | | |

The values of the model parameters[8] suggested in [35] are summarised in Table 3. Figure 11 shows some geometric properties of the function $K(x, y)$ for

---

[8] The model parameters are normalised, so the state variables in this section do not represent actual data.

the given parameters in the $(x, y)$ positive quadrant. The dotted line describes the set $K(x, y) = 0$. Since $K(x, y) > 0$ above this line, the immune term $w$ increases when $(x, y)$ belongs to this area. The dashed line and the solid line describe the sets $K(x, y) = 0.5$ and $K(x, y) = 1$, respectively. $\pi_A$, $\pi_B$, and $\pi_C$ correspond to the projection into the $(x, y)$-plane of Point A, Point B, and Point C, respectively. From Fig. 11 we can realise that if we force the patient state to be such that $\dot{w}(t) > 0$[9] then the immunity can be enhanced.

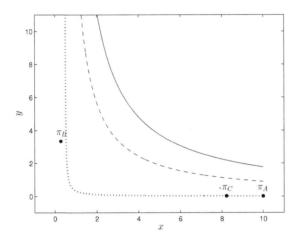

**Fig. 11.** Graphical visualisation of the immune system enhancing property in model (14-18). The dotted line indicates the set $K(x, y) = 0$. The dashed line and the solid line represent the sets $K(x, y) = 0.5$ and $K(x, y) = 1$, respectively. $\pi_A$, $\pi_B$, and $\pi_C$ correspond to the projection into the $(x, y)$-plane of Point A, Point B, and Point C, respectively.

### 4.3   Model Extension with Pharmacological Concepts

For most of the existing control strategies for HIV infection dynamics (e.g. [1,4,11]) the computed control input is hardly applied directly to real life of patients. This is because, instead of drug dosage, treatment efficacy is usually considered as the control input. The efficacy varies from 0 (no medication) to 1 (full medication), not expressed in terms of drug amounts. Now we address the relation between this control input and the real drug dosage in consideration of pharmacokinetic and pharmacodynamic models.

The process of drug administration is classically divided into two phases [30]: pharmacokinetic phase that relates dose, frequency, and route of administration to drug level-time course in the human body, and pharmacodynamic phase that relates the concentration of the drug at the sites of action to the magnitude of the produced effects.

---

[9] In [5,7,33] control of the drug dose is exploited so that the inequality is achieved.

**Pharmacokinetic Model.** Generally medicines are prescribed to be taken in a constant dosage at constant time intervals. In this study the drug is assumed to be absorbed completely and instantaneously. Also we assume that the drug distributes in an one-compartmental human body, and that it is eliminated from the human body by first order kinetics, in order for the simplicity of discussion (more sophisticated consideration can be a subject for future research direction). The rate of drug elimination is modelled by

$$\dot{\sigma} = -k\sigma, \tag{19}$$

where $\sigma$ is the intracellular amount of drug in the human body at time $t$ and $k$ is the constant elimination rate of the first order. Note that $k$ is related to the intracellular half-life of the drug $t_{\frac{1}{2}}$, i.e. $t_{\frac{1}{2}} = \log 2/k$. The intake of the HIV drug is considered as an impulsive input later in this study.

**Pharmacodynamic Model.** Numerous pharmacodynamic models have been investigated for fitting response-concentration curves empirically [12]. A class of these models can be described by the equation

$$\eta(t) = \eta_{max} \frac{C(t)^{\gamma}}{C(t)^{\gamma} + 1/Q}, \tag{20}$$

where $\eta(t)$ is the drug efficiency response at drug plasma concentration $C(t)$ and $\eta_{max}$ is the maximum response. $\gamma$ and $Q$ are constants.

As in [12] we consider $Q = 1/C_{50}$, where $C_{50}$ is the plasma concentration of drug which reduces the drug effect by 50 % of $\eta_{max}$. Based on [19,31,32] it is assumed that $\gamma = 1$, due to the fact that the FTC-TDF is a combination of emtricitabine and tenofovir, two nucleoside reverse transcriptase inhibitors (NRTIs).

In the pharmacokinetic model of Subsect. 4.3 we consider the human body as one-compartment. Then at time $t$ the amount of drug in the human body is the product of the plasma concentration $C(t)$, the apparent volume of distribution $V_d$, and the mass of the human body $M$. Hence $\sigma(t) = C(t)V_d M$, $\sigma_{50} = C_{50}V_d M$, and

$$\eta(t) = \eta_{max} \frac{\sigma(t)}{\sigma(t) + \sigma_{50}}. \tag{21}$$

For a full explanation of the pharmacological modelling, see [24].

**Pharmacological Model Parameters.** We consider the HIV drug FTC-TDF as in [15]. One FTC-TDF tablet is a combination of FTC (200 mg) and TDF (300 mg). To describe the effect of FTC-TDF we employ two pharmacological models for FTC and TDF, respectively.

In this subsection we suggest the pharmacological parameters $k_F = 0.0178$ $(/h)$ (i.e., 0.4266 $(/day)$), $k_T = 0.0042$ $(/h)$ (i.e., 0.1014 $(/day)$), $\sigma_{50,F} = 0.0069$ $(mg/l)$, and $\sigma_{50,T} = 0.3674$ $(mg/l)$ where the subscripts $F$ and $T$ stand for FTC and TDF, respectively. For the evaluation process of these parameters see [10].

As in [24] the average weight of a subject is assumed to be 70 $kg$. Then $\sigma_{50,F}$ is obtained as 0.6762 ($= 0.0069 \times 1.4 \times 70$), and $\sigma_{50,T}$ as 20.5744 ($= 0.3674 \times 0.8 \times 70$).

### 4.4 Integrated Model with Impulsive Input

The integrated model of (14-18) with pharmacological system for FTC and TDF (i.e. (19 and 21)) is given by

$$\dot{x} = \lambda - dx - (1 - \eta_F)(1 - \eta_T)\beta xy, \tag{22}$$

$$\dot{y} = (1 - \eta_F)(1 - \eta_T)\beta xy - ay - p_1 z_1 y - p_2 z_2 y, \tag{23}$$

$$\dot{z}_1 = c_1 z_1 y - b_1 z_1, \tag{24}$$

$$\dot{w} = c_2 xyw - c_2 qyw - b_2 w, \tag{25}$$

$$\dot{z}_2 = c_2 qyw - hz_2, \tag{26}$$

$$\dot{\sigma}_F = -k_F \sigma_F, \tag{27}$$

$$\dot{\sigma}_T = -k_T \sigma_T, \tag{28}$$

where

$$\eta_F = \eta_{max,F}\frac{\sigma_F}{\sigma_F + \sigma_{50,F}},$$

$$\eta_T = \eta_{max,T}\frac{\sigma_T}{\sigma_T + \sigma_{50,T}}.$$

In this model we assume that the antiretroviral effects of both drugs are independent of each other and the subscripts $F$ and $T$ stand for FTC and TDF, respectively. Let $X(t) := [x(t), y(t), z_1(t), w(t), z_2(t), \sigma_F(t), \sigma_T(t)]^T$ and represent model (22-28) by the equation

$$\dot{X} = F(X). \tag{29}$$

We now consider a conceptual realisation of the experiment of [15] via the model (29). To this end we should introduce impulsive inputs to the system: FTC and TDF intake once daily and possible exposure to HIV during the experiment. These correspond to impulsive changes of the states $\sigma_F$, $\sigma_T$, and $y$, respectively.

To describe approximately the impulsive inputs we employ a hybrid systems description as in [9],[10] hence the system with impulsive input can be described by

$$\dot{X}(t) = F(X(t)), \quad \text{if } t \notin S_d \cup S_v,$$

$$X^+ = X + V_d, \quad \text{if } t \in S_d,$$

$$X^+ = X + V_v, \quad \text{if } t \in S_v,$$

where $V_d$ is the vector corresponding to the magnitude of change of the states $\sigma_F$ and $\sigma_T$, while $V_v$ is of the state $y$. $S_d$ is the set of time instants in which the impulsive changes occur for the states $\sigma_F$ and $\sigma_T$, while $S_v$ is for the state $y$.

---

[10] Alternatively impulsive input can be modeled by a Dirac delta function as in [28].

To present the impulsive changes of $\sigma_F$ and $\sigma_T$ in the hybrid system, we introduce 'oral bioavailability' into $V_d$, a pharmacokinetic parameter describing the available fraction of an administered drug that reaches the system. Thus

$$V_d = [0, 0, 0, 0, 0, 200(mg) \times B_F, 300(mg) \times B_T]^T, \tag{30}$$

where $B_F$ and $B_T$ are the bioavailability of FTC and TDF, respectively. The median of $B_F$ is 0.92 while that of $B_T$ is 0.25 [13]. In addition we assume that once some fraction of the oral dose of FTC-TDF reaches the system, it is fully converted to the activated form intracellularly.

$S_d = \{1, 2, 3, \cdots, T_d\}$, with $T_d = 450$ (day), following the protocol of the experiment in [15]. The values of $M_v$ in $V_v = [0, M_v, 0, 0, 0, 0, 0]^T$ and the set $S_v$ are selected in this section.

### 4.5    Numerical Simulations

In this section we consider that the initial state $X(0)$ is $[10, 0, 10^{-4}, 10^{-4}, 10^{-4}, 0, 0]^T$, which represents the status of a HIV-free patient that corresponds to most of the subjects of the experiment in [15].

We assume $M_v = 10^{-4}$ and $S_v = \{10, 50\}$, which imply that possible contacts with HIV occur at the time instants of the set $S_v$ and a small number of infected cells $y$ emerges by the infusions of the virus into the human body. To follow the experiment protocol we consider that the contact with the virus happens only in the duration of the drug prescription by setting $\max(S_v) < T_d$ and that the drug administration is carried out at $t = 1, 2, 3, \cdots, T_d$.

**Two Cases with Different Pharmacological Parameters.** In this subsection we simply assume that $\eta_{max,F} = 1$ and $\eta_{max,T} = 1$ based on [19,31,32] since the drug FTC-TDF is a combination of two NRTIs. This implies that there is not any drug-resistant virus species particularly against FTC and TDF in the system. A different scenario is discussed in the next subsection.

*The Case with Representative Phamacological Parameters*

Figure 12 shows the results of the simulation of the experimental protocol in [15] (note the different time scales). We use the proposed hybrid system and the parameter values of $k_F$, $k_T$, $\sigma_{50,F}$, and $\sigma_{50,T}$, obtained in Subsect. 4.3. In the upper left graph of Fig. 12, the solid line and the dotted line indicate the time histories of $\sigma_F$ and $\sigma_T$, respectively.

The resulting $(x, y)$ trajectory is displayed in Fig. 13. The solid line indicates the $(x, y)$ trajectory. Note that the projection of the initial state is on $\pi_A$ and that the trajectory is placed in the region $K(x, y) < 0$.

For this simulation the $(x, y)$ point stays extremely close to $\pi_A$. Although the level of $w$ is not boosted (since the $(x, y)$ point is located in the region in which $K(x, y)$ is negative for the whole simulation time), the level of $y$ becomes less than $1.7577 \times 10^{-39}$ at the end of the simulation. This case might represent the HIV preventive case of [15].

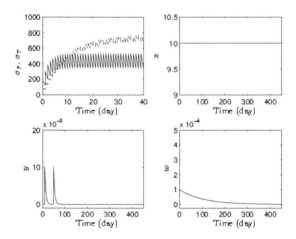

**Fig. 12.** Results of the numerical realisation of the experiment in [15] via the model (29) with impulsive inputs: $k_F = 0.4266$, $k_T = 0.1014$, $\sigma_{50,F} = 0.6762$, and $\sigma_{50,T} = 20.5744$. In the upper left graph, the solid line and the dotted line indicate the time histories of $\sigma_F$ and $\sigma_T$, respectively. For the implementation of the impulses we set $M_v = 10^{-4}$ and $S_v = \{10, 50\}$. Note the different time scales.

As pointed out in [25,37], our model is deterministic and the virus load ($y$) cannot be reduced to zero exactly. Note that cell and virus are countable biological object in the human body while the state variables of model (29) are continuous functions of time. Thus we can assume that HIV is eradicated if the virus load is reduced to a sufficiently low level. For example, in [25,37], virus elimination has been studied using a threshold of viral extinction, which could

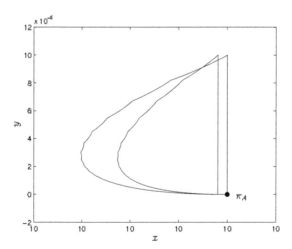

**Fig. 13.** The $(x, y)$ trajectory resulting from the simulation presented in Fig. 12. The projection of the initial state is on $\pi_A$. Note that the trajectory is placed in the region $K(x, y) < 0$.

correspond to a virus population less than one cell. We study viral clearance with a threshold of virus load in the next subsection.

*The Case with Comparatively Insensitive Response to the Drug*

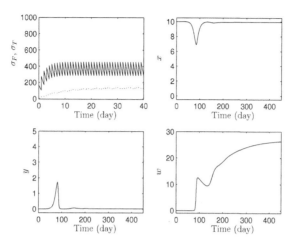

**Fig. 14.** Results of the numerical realisation of the experiment in [15] via the model (29) with impulsive inputs: $k_F = 0.4621$, $k_T = 0.1094$, $\sigma_{50,F} = 18.8020$, $\sigma_{50,T} = 302.4$, $B_F = 0.84$, and $B_T = 0.05$. These are the estimated values to present a subject showing insensitive drug response, based on the pharmacological parameter ranges in [13,18]. In the upper left graph, the solid line and the dotted line indicate the time histories of $\sigma_F$ and $\sigma_T$, respectively. Note the different time scales.

In Subsect. 4.3 we obtain the parameters ($k_F$, $k_T$, $\sigma_{50,F}$, and $\sigma_{50,T}$) using representative values such as median or geometric mean, for the pharmacological parameter ranges in [13,18]. In this subsection we consider the parameters for a subject showing relatively insensitive response to the HIV drug based on the given ranges of the pharmacological parameters, namely, $k_F = 0.0193$ /h (0.4621 /day), $k_T = 0.0046$ /h (0.1094 /day), $\sigma_{50,F} = 18.8020$ mg, $\sigma_{50,T} = 302.4$ mg, $B_F = 0.84$ and $B_T = 0.05$. See [10] for the procedure to obtain these parameters.

Simulation results are presented in Figs. 14 and 15. Figure 14 shows the results of the simulation with the newly estimated values of $k_F$, $k_T$, $\sigma_{50,F}$, $\sigma_{50,T}$, $B_F$, and $B_T$ (note the different time scales). In the upper left graph of Fig. 14, the solid line and the dotted line indicate the time histories of $\sigma_F$ and $\sigma_T$, respectively.

The resulting $(x, y)$ trajectory is displayed in Fig. 15. The dotted line indicates the set $K(x, y) = 0$ and the solid line indicates the $(x, y)$ trajectory. $\pi_A$, $\pi_B$, and $\pi_C$ are the projections onto the $x - y$ plane of Point A, Point B, and Point C, respectively. Note that the projection of the initial state is on $\pi_A$. The bottom graph is a zoomed-in version of the top graph.

Throughout the simulation the $(x, y)$ point is kept around $\pi_A$ and $\pi_C$ with comparatively high and low levels of the states $x$ and $y$, respectively, consistently

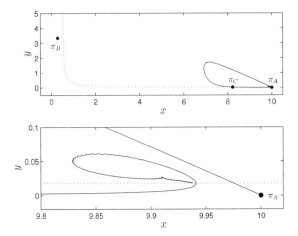

**Fig. 15.** The $(x, y)$ trajectory resulting from the simulation presented in Fig. 14. The dotted line indicates the set $K(x, y) = 0$ and the solid line indicates the $(x, y)$ trajectory. The projections of Point A, Point B, and Point C is on $\pi_A$, $\pi_B$, and $\pi_C$, respectively.

with the HIV preventive effect shown in [15]. Furthermore the level of $w$ is boosted over 25, as explained in Subsect. 4.2. Given the analysis in Subsect. 4.2 the $w$ state is enhanced when the $(x, y)$ point is located in the region in which $K(x, y)$ is positive. If the $(x, y)$ trajectory stays within the region $c_2 xy - c_2 qy - b_2 < 0$ (i.e. the region below the dotted line in Fig. 15), then the concentration of CTL precursor $w$ decreases. Although the $(x, y)$ point stays temporary within the region in which $K$ is negative, the state $w$ is boosted at the end of the simulation.

**Parameter Plane of** $\eta_{max,F}$ **and** $\eta_{max,T}$. In the previous subsection we ideally assume that drug-resistant virus does not exist for the two NRTI, FTC and TDF. In this subsection we study a different scenario. Note that anti-retroviral HIV treatment often fails due to the emergence of resistant virus [25].

The emergence of resistant virus results in the reduction of drug effect. Thus, to describe the effect of resistant virus on the drug treatment, we consider variation of the parameters, $\eta_{max,F}$ and $\eta_{max,T}$, which have been assumed equal to one in the previous simulation studies. All other parameters are as in Subsect. 4.5.

To exploit the range of $\eta_{max,F}$ and $\eta_{max,T}$ between 0 and 1, a parameter plane $(\eta_{max,F}, \eta_{max,T})$ is considered. We consider 31 evenly spaced parameter values along each axis of the plane. At every pair of parameter values (i.e. 31 × 31 pairs) we simulate with impulsive input for 450 days, as in Subsect. 4.5.

After the simulation with impulsive input we determine if the virus load $(y)$ is less than a predefined level of viral extinction threshold, $y_{EXT}$. If so, then we consider that this case could correspond to the HIV prevention case in the experiment of [15]. If not, then we investigate in which region of attraction, i.e. that of LTNP (Point C) or of AIDS (Point B), the final state of each simulation is located.

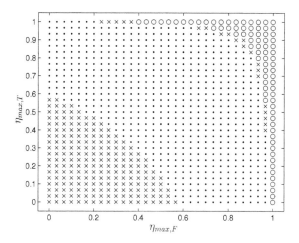

**Fig. 16.** The convergence results in the parameter plane $(\eta_{max,F}, \eta_{max,T})$. We consider 31 evenly spaced parameter values along each axis. After the simulation with impulsive input, if the virus load is less than viral extinction threshold $y_{EXT}$, then the corresponding point in the plane is marked with circle. Otherwise we investigate in which of the regions of attraction the final state is located. The dot and cross marks correspond to convergence to LTNP and AIDS, respectively.

To this end we additionally simulate for $2,000$ more days without any input. After this additional input-free simulation we measure the 2-norm distance between this new final state and Point C (or Point B). If the distance to Point C (or Point B) is less than a predefined bound, $B_{RA}$, then we regard that the state converges to Point C (or Point B). Otherwise we conclude that the state does not converge to Point C or B either.

The investigation results upon the parameter plane are presented in Fig. 16. For the study of this parameter plane, let $y_{EXT} = 10^{-10}$ and $B_{RA} = 0.1$. These results are plotted in the figure with the dot and cross marks, corresponding to convergence to LTNP and AIDS, respectively. The circle marks in the figure represents the case of HIV prevention.

As shown in the case with comparatively insensitive response to the drug, the protocol of the experiment could enhance the immune response against HIV for some subjects. Based on the simulation study beyond the drug administration (e.g. Fig. 16) we could suggest longer follow-up of the subject of the experiment after cessation of drug therapy in order to determine to which region of attraction the final state of each subject is located.

For the cases of convergence towards AIDS (Point B) we provide two explanation. If the drugs are too ineffective to suppress the virus, then the subject might develop full-blown AIDS (the lower left part of Fig. 16). If the state stays in the region $K(x, y) > 0$ and in the region $y > y_{EXT}$ for the duration of the drug treatment simulation, then the immune system is not boosted and also we cannot conclude that the virus is eradicated (the upper right part of Fig. 16).

## 5    Conclusions

We have investigated a control method to enhance the immune response in disease dynamics such as HIV and malaria. Then we have applied the control method to the Th1-Th2-pathogen population model to select the appropriate immune response between the Th1 and the Th2 responses. In this chapter the idea of hybrid control is used and we also propose physically available methods to implement the hybrid control in each biological system.

### 5.1    Further Remarks on Malaria Control

The experimental result in [29] related to single species malaria infection could be evidence to support the theoretical studies of open-loop control of the malaria dynamic model in this chapter. At the moment the malaria model used in the chapter can provide only a crude cartoon of the complex dynamics between participating components and reactions related to intruding parasite or pathogen [17].

### 5.2    Further Remarks on Venom Immunotherapy Control

The reason why the switch Th1-Th2 is guaranteed but the switch Th2-Th1 is not with the current parameters is found in Fig. 7. By the control method suggested in Subsect. 3.2, the state $[x_1(t), x_2(t), p(t)]^T$ is located in the region in which $M_2 \geq 0$ for $0 < t \leq T_t$, whereas the state could be placed in either $M_1 \geq 0$ or $M_1 < 0$ during the control treatment. When the state is in the region in which $M_1 < 0$ and $M_2 \geq 0$, it moves towards Point Th2, being away from Point Th1. If the state is in the region in which $M_1 \geq 0$ and $M_2 \geq 0$, then $x_1$ and $x_2$ are increased. This implies that the difference between $p$ values of the two surfaces $M_1 = 0$ and $M_2 = 0$ in Fig. 7 increases, and that the state moves into the region in which $M_1 < 0$ and $M_2 \geq 0$ in finite time by the proposed control scheme (note that in this chapter we use fixed size of state jumps). Thus due to Fig. 7 we are capable of driving the state towards Point Th2. For example if we have $\sigma_1 = 0.5$ and $\sigma_2 = 8$ then the switch Th2-Th1 is guaranteed instead of the switch Th1-Th2.

A switch from Th1 to Th2 is induced if antigen concentration is a sufficiently high value [3]. For example high antigen doses such as $p(0) = 50$ in model (10-12) induce a shift to the Th2 dominance status [2]. Note that in the proposed control method we use considerably small sizes of controlled jumps in the state $p$, namely 0.1 or 1.

The induction of Th1-Th2 switch is one of the mechanism of the selection of the appropriate T helper response [2]. By the proposed control method we can accelerate the onset of the Th2 response if it is the appropriate immune response.

## References

1. Adams, B.M., Banks, H.T., Kwon, H., Tran, H.T.: Dynamic multidrug therapies for HIV: optimal and STI control approaches. Math. Biosci. Eng. **1**, 223–241 (2004)

2. Bergmann, C., van Hemmen, J.L., Segel, L.E.: Th1 or Th2: How an appropriate T helper response can be made. Bull. Math. Biol. **63**, 405–430 (2001)
3. Bergmann, C., van Hemmen, J.L., Segel, L.E.: How instruction and feedback can select the appropriate T helper response. Bull. Math. Biol. **64**, 425–446 (2002)
4. Brandt, M.E., Chen, G.: Feedback control of a biodynamical model of HIV-1. IEEE Trans. Biomed. Eng. **48**(7), 754–759 (2001)
5. Chang, H., Astolfi, A.: Activation of immune response in disease dynamics via controlled drug scheduling. IEEE Trans. Autom. Sci. Eng. **6**, 248–255 (2009)
6. Chang, H., Astolfi, A.: Enhancement of the immune response to chronic myeloid leukaemia via controlled treatment scheduling. In: Proceedings of the of 31st Annual EMBS International Conference, pp. 3889–3892 (2009)
7. Chang, H., Astolfi, A.: Control of HIV infection dynamics: Approximating high-order dynamics by adapting reduced-order model parameters. IEEE Control Syst. Mag. **28**, 28–39 (2008)
8. Chang, H., Astolfi, A., Shim, H.: A control theoretic approach to venom immunotherapy with state jumps. In: 2010 Annual International Conference of the IEEE Engineering in Medicine and Biology Society (EMBC), pp. 742–745, August 2010
9. Chang, H., Astolfi, A., Shim, H.: A control theoretic approach to malaria immunotherapy with state jumps. Automatica **47**(6), 1271–1277 (2011). Special Issue on Systems Biology
10. Chang, H., Moog, C.H., Astolfi, A., Rivadeneira, P.S.: A control systems analysis of HIV prevention model using impulsive input. Biomed. Sign. Process. Control **13**, 123–131 (2014)
11. Ge, S.S., Tian, Z., Lee, T.H.: Nonlinear control of a dynamic model of HIV-1. IEEE Trans. Biomed. Eng. **52**(3), 353–361 (2005)
12. Gibaldi, M., Perrier, D.: Drugs and the Pharmaceutical Sciences, vol. 1: Pharmacokinetics. Marcel Dekker Inc, New York (1975)
13. Gilead Sciences Ltd., Truvada® Full Prescribing Information, 4 Dec 2011. Available at http://www.truvada.com/pdf/fpi.pdf. (2011)
14. David, B.K.: Golden. Insect sting allergy and venom immunotherapy: A model and a mystery. J. Allergy Clin. Immunol. **115**(3), 439–447 (2005)
15. Grant, R.M., Lama, J.R., Anderson, P.L., McMahan, V., Liu, A.Y., Vargas, L., Goicochea, P., et al.: Preexposure chemoprophylaxis for HIV prevention in men who have sex with men. New Engl. J. Med. **363**(27), 2587–2599 (2010)
16. Gurarie, D., McKenzie, F.E.: Dynamics of immune response and drug resistance in malaria infection. Malaria J. **5**(1), 86 (2006)
17. Gurarie, D., Zimmerman, P.A., King, C.H.: Dynamic regulation of single-and mixed-species malaria infection: insights to specific and non-specific mechanisms of control. J. Theor. Biol. **240**(2), 185–199 (2006)
18. Jackson, A., Moyle, G., Watson, V., Tjia, J., Ammara, A., Back, D., Mohabeer, M., Gazzard, B., Boffito, M.: Tenofovir, emtricitabine intracellular and plasma, and efavirenz plasma concentration decay following drug intake cessation: implications for HIV treatment and prevention. JAIDS, Publish Ahead of Print (2013)
19. Jilek, B.L., Zarr, M., Sampah, M.E., Rabi, S.A., Bullen, C.K., Lai, J., Shen, L., Siliciano, R.F.: A quantitative basis for antiretroviral therapy for HIV-1 infection. Nature Med. **18**, 446–451 (2012)
20. Kim, K.-W., Shin, Y.-S., Kim, K.-S., Chang, Y.-C., Park, K.-K., Park, J.-B., Choe, J.-Y., Lee, K.-G., Kang, M.-S., Park, Y.-G., Kim, C.-H.: Suppressive effects of bee venom on the immune responses in collagen-induced arthritis in rats. Phytomedicine **15**(12), 1099–1107 (2008)

21. Mason, D.P., McKenzie, F.E.: Blood-stage dynamics and clinical implications of mixed Plasmodium vivax-Plasmodium falciparum infections. Am. J. Trop Med. Hyg. **61**(3), 367–374 (1999)
22. Mason, D.P., McKenzie, F.E., Bossert, W.H.: The blood-stage dynamics of mixed Plasmodium malariae-Plasmodium falciparum infections. J. Theor. Biol. **198**(4), 549–566 (1999)
23. McQueen, P.G., McKenzie, F.E.: Host control of malaria infections: Constraints on immune and erythropoeitic response kinetics. PLoS Comput. Biol. **4**, e1000149 (2008)
24. Mhawej, M., Moog, C.H., Biafore, F., Brunet-François, C.: Control of the HIV infection and drug dosage. Biomed. Sign. Process. Control **5**(1), 45–52 (2010)
25. Nowak, M.A., May, R.M.: Virus Dynamics. Oxford University Press, New York (2000)
26. Recker, M., Nee, S., Bull, P.C., Kinyanjui, S., Marsh, K., Newbold, C., Gupta, S.: Transient cross-reactive immune responses can orchestrate antigenic variation in malaria. Nature **429**(6991), 555–558 (2004)
27. Richter, J., Metzner, G., Behn, U.: Mathematical modelling of venom immunotherapy. J. Theor. Med. **4**, 119–132 (2002)
28. Rivadeneira, P.S., Moog, C.H.: Impulsive control of single-input nonlinear systems with application to HIV dynamics. Appl. Math. Comput. **218**(17), 8462–8474 (2012)
29. Roestenberg, M., McCall, M., Hopman, J., et al.: Protection against a malaria challenge by sporozoite inoculation. N. Engl. J. Med. **361**(5), 468–477 (2009)
30. Rowland, M., Tozer, T.: Clinical Pharmacokinetics: Concepts and Applications. Lea & Febiger, Philadelphia (1980)
31. Sampah, M.E.S., Shen, L., Jilek, B.L., Siliciano, R.F.: Dose-response curve slope is a missing dimension in the analysis of HIV-1 drug resistance. Proc. Natl. Acad. Sci. **108**(18), 7613–7618 (2011)
32. Shen, L., Peterson, S., Sedaghat, A.R., McMahon, M.A., Callender, M., Zhang, H., Zhou, Y., Pitt, E., Anderson, K.S., Acosta, E.P., Siliciano, R.F.: Dose-response curve slope sets class-specific limits on inhibitory potential of anti-HIV drugs. Nature Med. **14**, 762–766 (2008)
33. Shim, H., Jo, N.H., Chang, H., Seo, J.H.: A system theoretic study on a treatment of AIDS patient by achieving long-term non-progressor. Automatica **45**, 611–622 (2009)
34. UNAIDS, T.W.H. Organization. AIDS epidemic update: 2010. UNAIDS, Geneva (2010)
35. Wodarz, D.: Helper-dependent vs. helper-independent CTL responses in HIV infection. J. Theor. Biol. **213**, 447–459 (2001)
36. Wodarz, D., Nowak, M.A.: Specific therapy regimes could lead to long-term immunological control of HIV. Proc. Natl. Acad. Sci. **96**(25), 14464–14469 (1999)
37. Wodarz, D., May, R.M., Nowak, M.A.: The role of antigen-independent persistence of memory cytotoxic T lymphocytes. Int. Immunol. **12**(4), 467–477 (2000)
38. World Health Organization Expert Committee on Malaria. 20th Report. WHO Regional Office for Africa (2003)
39. Zurakowski, R., Teel, A.R.: A model predictive control based scheduling method for HIV therapy. J. Theor. Biol. **238**, 368–382 (2006)

# Fluorescent Reporter Genes and the Analysis of Bacterial Regulatory Networks

Hidde de Jong[1]([⊠]) and Johannes Geiselmann[1,2]

[1] INRIA, Grenoble - Rhône-Alpes Research Centre, 655 avenue de l'Europe,
Montbonnot, 38334 Saint-Ismier Cedex, France
Hidde.de-Jong@inria.fr
[2] Laboratoire Interdisciplinaire de Physique (LIPhy, CNRS UMR 5588),
Université Grenoble Alpes, 140 Avenue de la Physique - BP 87,
38402 Saint Martin d'Hères, France
Hans.Geiselmann@ujf-grenoble.fr

**Abstract.** The understanding of the regulatory networks controlling the adaptation of bacteria to changes in their environment is critically dependent on the ability to monitor the dynamics of gene expression. Here, we review the use of fluorescent reporter genes for dynamically quantifying promoter activity and other quantities characterizing gene expression. We discuss critical physical and biological parameters in the design, development, and use of fluorescent reporter strains. Moreover, we review measurement models that have been proposed to interpret primary fluorescence data and inference methods for estimating gene expression profiles from these data. As an illustration of the use of fluorescent reporter strains for analyzing bacterial regulatory networks, we consider two applications in the model bacterium *Escherichia coli* in some detail: the joint control of gene expression by global physiological effects and specific regulatory interactions, and the importance of protein stability for the inference and analysis of transcriptional regulatory networks. We conclude by discussing some current trends in the use of fluorescent reporter genes.

**Keywords:** Fluorescent reporter genes · Bacterial regulatory networks · Growth · Gene expression · Bioinformatics · Systems biology

## 1 Introduction

Bacterial cells are capable of surviving in an enormous variety of conditions by adapting their functioning to changes in the environment. These adaptive capabilities emerge from complex regulatory networks that control and coordinate the different functions of the cell, including transport of substrates into the cell, the metabolism of these substrates to produce energy and molecular building blocks for growth, gene expression to convert the molecular building blocks into proteins, and the replication of DNA [40]. On the molecular level, these regulatory networks involve interactions between, for example, proteins and DNA

© Springer International Publishing Switzerland 2015
O. Maler et al. (Eds.): HSB 2013 and 2014, LNBI 7699, pp. 27–50, 2015.
DOI: 10.1007/978-3-319-27656-4_2

(transcriptional regulation) and between metabolites and proteins (transcription factor activity and enzyme activity). While for some model organisms the regulatory networks have been mapped in quite some detail, in most cases the structure of interactions is largely unknown. Moreover, even when the structure is known, the precise role of these interactions in bringing about a specific physiological response to a change in the environment remains a very difficult question for which only partial questions are available [2]. In order to reconstruct regulatory networks from experimental data, and understand the role they play in controlling the responses of bacteria to changes in their environment, we need to be able to dynamically quantify the metabolites, proteins, and mRNAs involved in cellular processes.

Since most of the mass contents of a cell are proteins and RNA, it is particularly important to measure gene expression over time. A large variety of technologies have been developed to this purpose, with particularly striking advances over the past decade. For example, DNA microarrays, RNA sequencing, and RT-qPCT technologies allow the quantification of cellular transcripts, often on a genome-wide scale [13,32]. Quantitative proteomics makes it possible to measure the protein contents of a cell [11], while ribosome profiling provides an estimate of the translation rates [20]. These technologies have provided a wealth of information on gene expression changes in bacteria and the regulatory networks that control these changes. However, for the purpose of reconstructing regulatory interactions from the data or observing precise temporal phenomena, they have limitations as well. For example, high-resolution time-series data are currently difficult to obtain, for both financial and technological reasons. Moreover, the above technologies involve many steps for extracting the cellular content and analyzing its constituents, thus potentially introducing biases. Furthermore, due to the requirement of having to remove a sample for analysis, these techniques do not allow monitoring the time course of gene expression in a single bacterial culture or in individual bacterial cells.

Fluorescent reporter genes provide an alternative and a complement to the above technologies, making it possible to monitor gene expression in real time and *in vivo*, both in single cells and on the population level. Reporter genes are classical tools in molecular biology, especially fusions of promoters and other regulatory regions with the *lacZ* gene. This gene encodes the protein $\beta$-galactosidase, the activity of which can be quantified by colorometric assays [42]. The breakthrough of fluorescent (or luminescent) reporter genes lies in combining the same principle of fusing a genomic region to a reporter gene with a non-intrusive technology for protein quantification [10,17,47]. When excited at a specific wavelength, the fluorescent protein encoded by the reporter gene produces light with an emission peak shifted toward the red with respect to the excitation wavelength. The emission peak can be easily captured and its intensity is proportional to the amount of fluorescent protein in the sample. The use of fluorescent reporter genes as tools in biology has enormously benefited from the engineering of new fluorescent proteins with desired spectral and other characteristics, molecular cloning techniques for constructing reporter systems inside the cell, and the miniaturization and automation of culturing bacterial cells and measuring emitted fluorescence. On the population level, microplate readers allow monitoring of the expression of fluorescent reporter genes in about one hundred bacterial cultures in parallel, whereas on the single-cell level advances in microfluidics and fluorescent microscopy make it possible to follow gene expression in single cells over many generations [29,49].

Reporter gene experiments generate huge amounts of data, the analysis of which poses subtle problems that are aften not straightforward to solve. The major difficulty in the analysis of the results of reporter gene experiments, usually consisting of emitted fluorescent levels and the optical density or absorbance of a growing culture (in the case of population-level measurements), lies in the indirect relation between the primary data and the biological quantities of interest. Making sense of the primary data requires mathematical models of reporter gene expression and statistical techniques to estimate the quantities in the models from the data [1,3,12,15,26,28,37,38,48,51]. Moreover, computer tools for the efficient and user-friendly application of these methods are needed, in order to make the analysis methods available to biologists [1,7,51].

The aim of this review is to give a short overview of the process of obtaining reliable data about gene expression by means of fluorescent reporter genes and to illustrate the use of the resulting data with some representative examples from the literature, in particular work carried out by the authors over the past few years. We will first discuss the physical and biological principles underlying fluorescent reporter genes and outline which experimental set-ups are typically used to follow bacterial gene expression over time. Our focus will be on population-level measurements of gene expression and we refer to other reviews for specific issues involved in single-cell experiments [36,49]. Second, we present different aspects of the analysis of the primary data thus obtained, in particular the measurement models used to interpret the primary data, the statistical methods to infer quantities in the model from the data, and computer tools to apply these methods in a rigorous and automated way. Third, we discuss how the use of fluorescent reporter genes has shed new light on a number of well-known model systems, such as the regulatory circuits involved in growth transitions in *E. coli* and chemotaxis and motility in the same organism.

## 2   Fluorescent Reporter Gene Experiments in Bacteria

### 2.1   Reporter Genes

Reporter genes are a very convenient and widely-used tool for measuring the dynamics of gene expression. The principle consists in duplicating the elements that control the expression of the gene of interest upstream of a readily detectable gene, the reporter gene. Depending on the control elements studied, different portions of the original gene are cloned upstream of the reporter gene [44]. For example, if we want to measure the combined effect of transcriptional and translational regulation, we would "fuse" the reporter protein to the gene of interest, creating a chimeric protein. This type of construction is called a translational fusion. Even though the basic principles are identical for all types of reporter gene constructions, we will focus here on the first, and most highly regulated step of gene expression: the initiation of transcription [8,40]. In order to study regulation of this process, we create "transcriptional fusions".

The fundamental principle is illustrated in Fig. 1. The elements that control transcription initiation are located in the promoter of the gene (red in Fig. 1). The time-dependent activity of the promoter determines the rate of production of the corresponding messenger RNA (mRNA), which is subsequently translated into protein. However, in almost all cases, the gene of interest codes for a protein that is not easily detectable. We therefore introduce into the cell a second copy of the promoter, cloned upstream of a gene coding for an easily detectable protein. This second copy can either be placed on the chromosome or carried on a plasmid [41]. The advantage of integrating the reporter construct into the chromosome is to ensure that in all conditions the number of copies of the reporter gene are well controlled. This criterion is important, for example, when using reporter genes for assessing the noise of gene expression in cells. However, extra-chromosomal copies carried on plasmids are easier to construct and give higher signal intensities due to the multiple copies of the plasmids. For the promoter of *E. coli*, a complete library fusing all intergenic regions to a *gfp* reporter gene have been constructed [50]. The copy-number of these plasmids remains constant over different growth conditions [6], thus avoiding the major potential artifact of the use of reporter plasmids.

## 2.2  Different Types of Reporter Genes

The most commonly used reporter proteins are fluorescent proteins [17,44]. These proteins contain a fluorophore that emits visible light shifted towards longer wavelengths when irradiated with light corresponding to excitation wavelengths of the molecule. The original green fluorescent protein (GFP) had an excitation maximum around 480 nm and an emission peak around 510 nm [47]. Today, variants of the protein are available that emit at virtually all visible wavelengths [10]. Since very few cellular components emit fluorescence in the visible part of the spectrum, the concentration of these proteins can be quantified within the living cell.

However, the formation of the fluorophore can be a relatively slow enzymatic reaction and the final step of the reaction sequence involves molecular oxygen [47]. These fluorescent reporters will therefore not function in an anaerobic environment. The engineered fluorescent proteins used today have maturation times on the order of 15 min, short compared to the dynamics of gene expression with characteristic times on the order of an hour. However, the bias of maturation rates should be taken into account and measured for each fluorescent protein and strain since this parameter can vary depending on experimental conditions [18]. When the maturation of the active fluorophore is incorporated into the measurement model, we can deduce the production rate of GFP from the experimentally observed fluorescence (see Sect. 3.1). Additional care must be taken when creating translational fusions since the attachment of the fluorescent protein to the

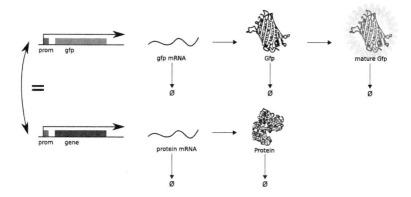

**Fig. 1. Reporter gene measurements.** The target gene (blue) is transcribed from the promoter (red, prom) to give the corresponding messenger RNA (mRNA). The mRNA is translated into the corresponding protein. However, the target protein is generally not easily detectable. The promoter of the target gene is therefore cloned upstream of the gene coding for the green fluorescent protein (green, GFP). Since the promoter region contains all of the control elements of transcription initiation, the rate of production of the *gfp* mRNA is in principle the same as the rate of production of the mRNA of the target gene. The *gfp* mRNA is translated into GFP, which is subsequently converted to the mature, fluorescent form by an autocatalytic process. All constituents of the process are degraded by growth dilution or natural degradation (Color figure online).

native protein can perturb the folding of GFP. Specific variants of GFP greatly alleviate this problem [46]. Even though the fluorescence of GFP allows the specific detection of this protein in a living cell, other cellular components, such as flavins, also emit fluorescence at similar wavelengths [28]. This autofluorescence has to be incorporated into the measurement model (Sect. 3.1).

Another type of reporter that gives a unique signal in living cells is luciferase. Most bacteria do not spontaneously emit light and bioluminescence is therefore an ultra-sensitive method for detecting the reporter protein in a bacterial population. The commonly used firefly luciferase is a monomeric protein that emits light when oxidizing the substrate, luciferin [39]. Bacterial luciferase is a heterodimer that emits light when oxidizing the substrate, a long-chain aldehyde [35]. Both of these enzymes can therefore not function in an anaerobic environment. Furthermore, the substrate of the bioluminescence reaction has to be continuously provided. This is difficult in the case of luciferin, but can be easily accomplished in the case of bacterial luciferase. We simply clone the entire luciferase operon, comprising genes *luxCDABE*, downstream of the promoter of interest. The genes *luxAB* code for the luciferase, whereas the genes *luxCDE* reduce cellular fatty acids to the aldehyde substrate of luciferase [35]. Reporter plasmids based on these constructs are readily available [33,34]. Light emission is therefore continuous. However, light emission depends on the concentration of the substrate and the concentration of luciferase. This dependency can be calibrated to derive a reliable measure of the concentration of luciferase from the measured bioluminescence intensity [33]. However, for both types of luciferases, the reaction depends on ATP and changes in the bioluminescence signal may reflect changes in intracellular metabolites. In the case of bacterial luciferase, the generation of the substrate is furthermore sensitive to the redox potential of the cell [25]. Even though bioluminescence detection is extremely sensitive, for all these reasons, fluorescent proteins are generally preferred as reporter genes.

## 2.3   Experimental Setup

A typical experiment of measuring the dynamics of gene expression therefore consists in constructing the appropriate plasmid-based or chromosomally integrated reporter construct (Sect. 2.1) and quantifying the fluorescence per cell as a function of time. The measurements can target single cells under the microscope [36, 49] or an entire bacterial population. In single-cell measurements the total fluorescence of the cell is divided by the cell volume in order to obtain the fluorescence intensity per unit cell volume. Here, we will focus on population measurements, where the total cell volume is approximated by the absorbance of the culture. The task therefore consists in measuring at regular time intervals the absorbance and the total fluorescence of the bacterial culture. The most convenient setup for this task consists of an automated, thermostated plate-reader that keeps the samples at a constant growth temperature and agitates the microplate in order to prevent sedimentation of the bacteria and provide sufficient mixing of the culture (Fig. 2). Typically, a 96-well microplate is agitated for about one minute, the absorbance signal of all wells is read, agitation is resumed and the fluorescence is measured. The entire growth and measurement cycle takes about 5 min, which is a sufficient sampling rate since changes in gene expression take place on a time-scale of several tens of minutes or even hours.

The goal of such experiments is to monitor the dynamics of gene expression. This implies that the environmental conditions change during the experiment. This is achieved by carefully selecting the initial conditions. For example, the bacteria could be grown overnight in an Erlenmeyer flask until they have exhausted the available nutrient source (Fig. 2). The experiment is started by diluting this pre-culture into fresh growth medium. The changes in gene expression then reflect the dynamics of adaptation of the bacteria to the new environment. In the example of Fig. 2, after adapting to new environmental conditions the bacteria reach a steady state where gene expression remains constant. In this example, the new growth medium contains two nutrient sources. When the preferred nutrient source is again exhausted, the bacteria continue growth, at a lower rate, on the less preferred nutrient and the expression of the target gene dynamically adapts to the new environment. A typical experiment of this type lasts for about 12 h. In principle, 96 different reporter gene constructs can be measured in parallel in a 96-well microplate. However, in order to analyze the data, we need to include wells of bacteria that do not express the fluorescent protein (for fluorescence background correction) and wells that do not contain bacteria in order to correct the measured absorbance for the absorbance of the

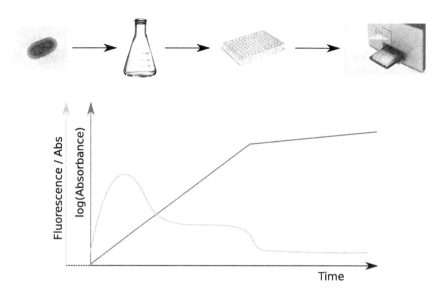

**Fig. 2. A typical experiment for measuring the dynamics of gene expression using reporter genes and an automated microplate reader.** The top row shows the main steps of the experiment. Bacteria are generally grown in a culture tube or an Erlenmeyer flask to a defined starting state. Aliquots of different pre-cultures are dispatched into the wells of a microtiter plate. The microplate is then read in an automated plate-reader that measures the absorbance and fluorescence signals of all the wells at regular intervals (typically five minutes). An example curve is shown in the bottom part of the figure. Here, the bacterial culture grows exponentially until a preferred nutrient source is exhausted. The bacteria continue to grow at a lower rate on a less favorable nutrient source. The absorbance measurements (blue) are on a logarithmic scale. The fluorescence measurements are shown in red on a linear scale. For clarity, we show the relative fluorescence, *i.e.*, the absolute fluorescence measured by the plate-reader divided by the absorbance (Color figure online).

growth medium (Sect. 3). It is even possible to incorporate two or three different reporter gene constructs, using fluorescent proteins of different color, into the same bacterial cell. In this case, each measurement cycle comprises fluorescence measurements at different wavelengths. An experiment thus generates a large amount of data that have be analyzed in an automated manner (Sect. 3.3).

# 3    Analysis of Fluorescent Reporter Gene Data

## 3.1    Mathematical Models of Reporter Gene Expression

The interpretation of reporter gene data is usually based on mathematical models describing the dynamics of the expression of reporter genes. The most commonly found models in the literature distinguish between the stages of transcription, translation, and maturation, giving rise to a three-variable system of ordinary differential equations (ODEs). Here, we follow with some modifications the model of [12]. The variables are $m(t)$, $r_u(t)$, and $r(t)$, representing the time-varying concentrations of reporter mRNA and immature and mature reporter protein, respectively ($\mu$M), with time variable $t$ (min).

The accumulation of reporter mRNA involves a balance between synthesis and decay. mRNA synthesis is defined as the product of a maximum synthesis constant $k_m$ ($\mu$M min$^{-1}$) and a dimensionless promoter activity $a(t)$, scaled between 0 and 1. Notice that the promoter activity is a time-dependent input variable and that the regulatory interactions shaping the temporal variation of the promoter activity are not explicitly modeled. In fact, as will be seen below, the very objective of reporter gene experiments is usually to infer $a(t)$ from the data. mRNA decay involves a first-order term representing physical degradation, with a rate constant $d_m$ (min$^{-1}$) and growth dilution, determined by the time-varying growth rate $\mu(t)$ (min$^{-1}$). In practice, the growth dilution term is often omitted since mRNA decay is dominated by physical degradation: mRNA in bacteria is generally unstable, with half-lives on the order of a few minutes [5], much shorter than typical cell doubling times. The above considerations give rise to the following equation describing the transcription of a gene:

$$\frac{d}{dt}m(t) = k_m\, a(t) - (d_m + \mu(t))\, m(t). \tag{1}$$

The dynamics of the concentration of unfolded (immature) protein is also defined as a balance between protein synthesis and decay. The synthesis rate is given by a first-order term, the multiplication of the reporter mRNA concentration $m(t)$ with a protein synthesis constant $k_u$ (min$^{-1}$), which can be interpreted as the translational activity (quantity of protein produced per quantity of mRNA per min). Here we will assume that $k_u$ is constant [24]. Note that this assumption implies that there is no (active) regulation of translation. The decay of immature protein also involves physical degradation and growth dilution, like for mRNA, with the degradation constant $d_r$ (min$^{-1}$). The half-live of a protein is defined in terms of the latter constant as $\ln 2/d_r$. In addition, the maturation of the unfolded protein leads to another first-order decay term, involving the maturation constant $k_r$ (min$^{-1}$). Maturation times reported in the literature are usually given as the half-time of maturation, $\ln 2/k_r$. The translation equation thus becomes

$$\frac{d}{dt}r_u(t) = k_u\, m(t) - (d_r + k_r + \mu(t))\, r_u(t). \tag{2}$$

Mature protein accumulates through maturation and decays by physical degradation and growth dilution, where we assume that the mature and immature proteins have the same half-lives and thus degradation constants:

$$\frac{d}{dt}r(t) = k_r\, r_u(t) - (d_r + \mu(t))\, r(t). \tag{3}$$

For a more detailed discussion of the assumptions underlying the above models, see [12]. Notice that the total concentration of reporter protein, $r_{tot}(t)$, can be obtained by summing Eqs. 2 and 3, giving rise to $d/dt\, r_{tot}(t) = k_u\, m(t) - (d_r + \mu(t))\, r_{tot}(t)$.

In practice, it is often possible to reduce the above models, in particular when the maturation and mRNA degradation rates are quite fast with respect to the growth rate or the protein degradation rate. This allows a quasi-steady state assumption (QSSA) to be applied, implying $d/dt\, m(t) = d/dt\, r_u(t) = 0$, and reducing the three-variable model to the following ODE:

$$\frac{d}{dt}r(t) = k'_m\, a(t) - (d_r + \mu(t))\, r(t), \tag{4}$$

where $k'_m$ (min$^{-1}$) is a lumped protein synthesis constant, defined as

$$k'_m = \frac{k_m\, k_u}{d_m}, \tag{5}$$

and where $k'_m\, a(t)$ defines the protein synthesis rate. Remark that in this case, the protein synthesis rate is directly proportional to the promoter activity. If the reporter protein is very stable and the bacterial cultures are growing at a non-negligible rate, we have $d_r \ll \mu(t)$ and the model can be further simplified by ignoring the term describing the physical degradation of the protein. As will be seen below, the empirical expression for promoter activity often found in the literature corresponds to the latter case.

## 3.2   Estimation of Gene Expression Profiles from Primary Data

Which quantities of biological interest can be derived from the optical density/absorbance and fluorescence data by means of the models introduced above? By strain construction, as explained in Sect. 2, the promoter activities of the gene of interest and the reporter gene are identical in transcriptional fusions. In [12,51] it is shown that, if the expression of the gene of interest is adequately described by the model structure of Eqs. 1–3, in particular when assuming constant translational activity (no regulation of translation) and constant half-lives of mRNA and protein (no regulation of mRNA and protein degradation), then the reporter gene model can be used to reconstruct the promoter activity $a(t)$ of the gene of interest, up to an unknown multiplicative coefficient. In the case of active regulation of translation and degradation, transcriptional fusions may not be sufficient to obtain precise estimates and the chromosomal regions involved in post-transcriptional regulation need to be integrated into the reporter construction (Sect. 2). In addition to promoter activities, other quantities of interest can be estimated from the data using the above model and some additional assumptions and extensions, e.g., protein concentrations [12,51] and concentrations of transcriptional regulators [15,19]. In the remainder of this section, for the sake of simplicity, we will focus on the estimation of promoter activities using the reduced model of Eq. 4.

Figure 3A shows an example of typical time-course fluorescence and absorbance data acquired in a reporter gene experiment, derived from [43].

An *E. coli* strain carrying a low-copy reporter plasmid with a transcriptional fusion of the promoter of the gene *tar*, encoding a chemoreceptor protein playing a role in motility. The strain, carrying a deletion of the transcription regulator CpxR, was grown in a thermostated and agitated microplate in minimal medium with glucose for almost 10 h, with a first phase of exponential growth followed by growth arrest after glucose exhaustion. The reporter protein is a stable and fast-maturing GFP, called GFPmut2 [50]. The microplate reader captures fluorescence and absorbance signals at regular time intervals, leading to more than 100 measurements over the time window of the experiment.

A first step in data analysis is the detection of outliers, not necessary in the example of Fig. 3, and the subtraction of background levels of fluorescence and absorbance. The corrected absorbance signal $A(t)$ is computed as

$$A(t) = A_u(t) - A_b(t), \tag{6}$$

where $A_u(t)$ is the primary absorbance signal and $A_b(t)$ the absorbance of the growth medium (M9 in this example). As explained in Sect. 2.3, the fluorescence signal needs to be corrected for autofluorescence generated by wild-type bacteria carrying a non-functional reporter plasmid or no plasmid at all (in practice these two measures of autofluorescence usually give the same result). Contrary to the absorbance background, the autofluorescence depends on the (time-varying) population size. An obvious strategy would be to directly subtract the autofluorescence of the control culture from the fluorescence of the culture of the bacteria carrying the reporter plasmid. This does not always work though, since the two cultures may not be exactly synchronized. As an alternative, one can use a calibration procedure, such that the corrected signal $I(t)$ is defined by

$$I(t) = I_u(t) - s(A(t)), \tag{7}$$

where $I_u(t)$ is the primary fluorescence level and $s$ a calibration function, mapping absorbance levels to autofluorescence levels [43]. The calibration function is obtained by fitting a cubic smoothing spline to the autofluorescence generated by bacteria carrying the non-functional reporter plasmid or no plasmid at all as a function of the absorbance. Splines have the advantage that they can be evaluated for any absorbance within the observed range and easily extrapolated beyond this range. Figure 3B-C gives an example of this strategy of background correction of absorbance and fluorescence data. Another strategy, in principle more powerful than the one used above, would be to measure autofluorescence at a different wavelength and use spectral unmixing to correct autofluorescence [28].

A second step is the computation of the promoter activity from the corrected absorbance and fluorescence signals. From Eq. 5 it follows that

$$k'_m a(t) = \frac{d}{dt} r(t) + (d_r + \mu(t)) r(t). \tag{8}$$

The growth rate $\mu(t)$ can be directly estimated from the absorbance, that is,

$$\mu(t) = \frac{d}{dt} A(t) \frac{1}{A(t)} = \frac{d \ln A(t)}{dt}. \tag{9}$$

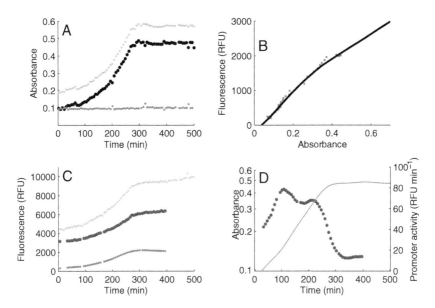

**Fig. 3. Illustration of the data analysis procedures** (adapted from [43]). Absorbance and fluorescence data acquired for the $\Delta cpxR$ mutant strain carrying a pUA66*tar-gfp* reporter plasmid, grown in M9 with glucose. *A:* Primary (uncorrected) absorbance (•, grey), background absorbance (•, red), and corrected absorbance (•, black). *B:* Calibration curve obtained by measuring the autofluorescence of the wild-type strain without plasmid. Primary fluorescence data are plotted against (corrected) absorbance data and the curve is obtained by fitting a smoothing spline. *C:* Primary fluorescence data (•, grey), and the corrected fluorescence (•, blue) obtained after subtracting the fluorescence of the background (•, red) as in Eq. 7. *D:* Promoter activity of *tar* (•, blue) computed from the corrected absorbance (-, grey) and corrected fluorescence by means of Eq. 11 (Color figure online).

The time-varying GFP concentration in the bacterial population, $r(t)$, can also be estimated from the absorbance and fluorescence, making the usual assumptions that the fluorescence is proportional to the number of GFP molecules and the absorbance proportional to the biomass (Sect. 2.3):

$$r(t) \sim \frac{I(t)}{A(t)}. \tag{10}$$

We arbitrarily set the proportionality constant in Eq. 10 to 1, thus expressing the reporter protein concentration in relative fluorescence units, RFU (and the synthesis rate in units RFU $min^{-1}$). Notice that this results in a relative, not an absolute quantification of promoter activity, as is common in the literature. In order to obtain an absolute quantification of promoter activity, an additional calibration step would be necessary to relate fluorescence units to the number of GFP molecules [45].

Substituting the expressions for $r(t)$ and $\mu(t)$ into Eq. 8 yields [12]:

$$k'_m \, a(t) = \frac{dI(t)}{dt} \frac{1}{A(t)} - \frac{dA(t)}{dt} \frac{I(t)}{A(t)^2} + \left( \frac{dA(t)}{dt} \frac{1}{A(t)} + d_r \right) \frac{I(t)}{A(t)}$$

$$= \frac{\frac{d}{dt}I(t)}{A(t)} + d_r \frac{I(t)}{A(t)}. \tag{11}$$

This definition is equivalent to other definitions in the literature [38] when $\mu(t) \gg d_r$. The expression can be evaluated using estimates of $A(t)$, $I(t)$, and $dI(t)/dt$ obtained by means of cubic smoothing splines, following the procedure in [12]. Moreover, the half-life of the fluorescent reporter, and thus the degradation constant $d_r$, can be easily measured experimentally. Figure 3D shows the promoter activity of *tar* computed from the data in panels *A* and *C*.

The approach outlined above is indirect, in the sense that it smooths the data first and reconstructs the promoter activity from the measurement model only in a second step. This results in the propagation of estimation errors that may be difficult to control. Other methods formulate a regularized data fitting problem directly in terms of the quantities to be estimated, thus solving the inference problem in a single and better controlled optimization step. Examples of the latter approach are linear inversion methods [3,51] and methods based on the use of Kalman filters [1].

### 3.3    Computer Tools for Analyzing Fluorescent Reporter Gene Data

Population-level experiments with reporter genes in microplate readers generate a huge amount of data, on the order of $10^4$–$10^5$ data points per microplate. In order to treat these data in a systematic and efficient way, and make the methods available to biologists, the latter need to be implemented and packaged in user-friendly computer tools. An example of such a tool is Wellreader, a Matlab program for the analysis of fluorescent and luminescent reporter gene data based on the indirect methods outlined above [7]. Other examples are WellInverter [51] and BasyLICA [1], which employ indirect approaches for inferring gene expression profiles. WellInverter is a web server application that provides a graphical user interface allowing online access to the linear inversion methods through a web-based platform. The user can upload experimental files by means of WellInverter, remove outliers and subtract background, and launch the procedures for computing growth rates, promoter activities, and protein concentrations. The methods underlying WellInverter are also available as a stand-alone Python package (WellFARE).

## 4    Use of Fluorescent Reporter Genes for Studying Bacterial Regulatory Networks

Reporter gene experiments in microplate readers contain information on population-level gene expression that is highly valuable for studying the adaptation of bacterial gene expression to changes in the environment. In this section, we provide two examples illustrating the use of such fluorescent reporter gene data sets.

## 4.1   Joint Control of Gene Expression Changes by Global Physiological Effects and Specific Regulatory Interactions

Regulatory networks controlling adaptation of gene expression in bacteria involve transcription factors that sense environmental and metabolic signals and specifically activate or inhibit target genes. In addition to such specific factors, gene expression also responds to changes in a variety of physiological parameters that modulate the rate of transcription and translation, such as the concentrations of (free) RNA polymerase and ribosome, gene copy number, and the size of amino acid and nucleotide pools. Contrary to specific regulators, these so-called global physiological effects affect the expression of all genes. The importance of global physiological effects, in particular the activity of the transcriptional and translational machinery in the control of gene expression, have been quantitatively assessed in recent publications [6, 16, 22, 24].

It is to be expected that, during the transition from one growth condition to another, both global physiological effects and specific regulatory interactions play a role in reorganizing the gene expression state of the bacterial cell. However, only very few studies have investigated the relative contributions of these factors in an actual regulatory network and dynamically, during a growth transition. Here, we summarize one such study carried out in our group [6], where we considered this question in the context of a central regulatory circuit of carbon metabolism in *E. coli*. The network, shown in Fig. 4*A*, consists of two transcription regulators, Crp and Fis, that regulate the expression of a large number of genes encoding enzymes in central metabolism, including the gene *acs*. Since the latter gene is strongly expressed in the absence of glucose, it provides an excellent indicator of the transcriptional response of carbon metabolism to a change in glucose availability and the accompanying change in growth rate. The signaling metabolite cyclic AMP (cAMP), which is required to activate Crp, also responds to a change in glucose availability by a strong increase of its intracellular concentration. Moreover, the expression of all genes is controlled by the activity of the transcriptional and translational machinery.

The original question was rephrased in the context of this network as follows. How do global physiological effects and the dense pattern of transcriptional regulatory interactions in Fig. 4*A* jointly contribute to the change in promoter activity observed during a growth switch? In particular, we considered the transition of *E. coli* cells growing in minimal medium supplemented with glucose as the sole carbon source to a complete arrest of growth due to the exhaustion of glucose. The cell maintains a minimal metabolism after glucose exhaustion by utilizing the low-energy substrate acetate that has been excreted during fast growth on glucose. During this transition between fast growth on glucose and slow growth on acetate, we measured the growth rate, the concentration of cAMP in the growth medium, from which we estimated the intracellular cAMP concentration, and the activity of the *acs*, *crp*, and *fis* promoters by means of plasmid-borne fluorescent reporter genes. In addition, in order to quantify the global physiological state we monitored the activity of a constitutively expressed promoter, that is, a promoter whose transcriptional activity only depends on the global physiological state of the cell, in particular the activity of the gene expression machinery [27]. The particular promoter used for this purpose was the $p_{RM}$ promoter, a phage promoter that is not known to be regulated by any specific transcription factor in non-infected *E. coli* cells.

In order to interpret the data obtained in these experiments, we proposed the following simple model of promoter activity. We denote by $p(t)$ the time-varying promoter activity [$\mu$M min$^{-1}$] and write

$$p(t) = k\, p_1(t)\, p_2(t), \tag{12}$$

where $k$ [$\mu$M min$^{-1}$] represents the maximum promoter activity. The dimensionless term $p_1(t)$, for convenience assumed to vary between 0 and 1, quantifies the modulation of the promoter activity by global physiological effects, for instance through the availability of free RNA polymerase. The dimensionless term $p_2(t)$, also varying between 0 and 1, accounts for the effect of transcription factors and other specific regulators. The unknown constant $k$ can be eliminated by normalizing Eq. 12 with respect to a reference state at time $t^0$, for instance steady-state growth on glucose. We define $p^0 = p(t^0)$, $p_1^0 = p_1(t^0)$, and $p_2^0 = p_2(t^0)$, and divide Eq. 12 by $p^0 = k\, p_1^0\, p_2^0$. A logarithmic transformation of the resulting terms results in

$$\log\frac{p(t)}{p^0} = \log\frac{p_1(t)}{p_1^0} + \log\frac{p_2(t)}{p_2^0}. \tag{13}$$

How can this very simple model be used to answer the question asked above? We considered two extreme cases, namely one in which it is assumed that specific regulators do not play any role in the dynamic regulation of the promoter activity. In this case, the second term in the right-hand side of Eq. 13 drops out of the equation and the model reduces to a simple equality of two terms. For instance, if we are interested in the promoter activity of the gene *crp*, we obtain $\log(p_{crp}(t)/p_{crp}^0) = \log(p_{RM}(t)/p_{RM}^0)$, where as stated above, the activity of the $p_{RM}$ promoter is used as a proxy for the quantification of global physiological effects. This hypothesis can be directly tested with the experimental data by plotting the normalized activities of the *crp* and $p_{RM}$ promoters in a scatter plot (Fig. 4*B*). If the assumption is correct and global physiological effects dominate regulation of *crp* activity, then the data points are expected to cluster around the diagonal, which is indeed seen to be the case ($R^2 = 0.96$). The same conclusion was reached for the *fis* promoter (data not shown), but in the case of the *acs* promoter the hypothesis clearly breaks down (Fig. 4*C*). However, we showed that a slightly adapted model fits the data well. In particular, when assuming that the specific regulatory effect is mostly determined by the concentration of cAMP, we obtain the following expression for the *acs* promoter from Eq. 13: $\log(p_{acs}(t)/p_{acs}^0) = \log(p_{RM}(t)/p_{RM}^0) + \log(c(t)/c^0)$, where $c$ represents the intracellular cAMP concentration. Since this concentration has also been measured, the model can be directly tested [6].

We have shown by means of the above analysis, and similar analyses performed in different growth conditions, with mutants strains, and for an additional target gene coding for a transcriptional regulator (RpoS), that the gene expression profiles can be adequately captured by the much simpler regulatory network shown in Fig. 4B. It can be directly seen that in comparison with the original network in panel A, apart from the transcriptional regulation of *acs* by the complex of Crp and cAMP, all specific regulatory interactions have disappeared. In other words, the assumption that global physiological effects dominate during the adaptation of *E. coli* cells to the exhaustion of glucose is sufficient to account for the observed gene expression profiles. While the control of gene expression during growth transitions is shared between global physiological effects and specific transcription factors, our results question the central role often attributed to transcriptional regulatory networks in controlling genome-wide expression changes during physiological transitions. It may be more appropriate to regard transcriptional regulators as complementing and finetuning the global control exerted by the physiological state of the cell.

Several other recent studies have used fluorescent reporter genes to study the shared control of gene expression in microorganisms by transcription factors and global cell physiology. For instance, Gerosa *et al.* have developed quantitative models to dissect global and specific regulation of *E. coli* genes involved in arginine biosynthesis [16]. Keren *et al.* have measured activities of 900 *S. cerevisiae* and 1800 *E. coli* promoters using fluorescent reporters. They showed that in both organisms 60 to 90 % of promoters change their expression between conditions by a constant global scaling factor that depends only on the conditions and not on the specific promoter considered, thus also suggesting that global physiological effects are an important driver for the adaptation of gene expression during growth transitions.

## 4.2 Importance of Protein Stability for Inference and Analysis of Transcriptional Regulatory Networks

The synthesis of flagella and the chemotaxis sensing system, enabling *E. coli* bacteria to orient themselves along gradients of certain chemicals in their environment, is under the control of a complex regulatory network. The more than 60 genes responsible for motility in bacteria are structured in a transcriptional hierarchy of three operon classes [9, 21, 31]. Several studies have used fluorescent reporter genes to better understand the functional organization of this hierarchy. For instance, Kalir et al. [21] found a detailed temporal ordering of the activation of the different promoters in the hierarchy, suggesting that flagella proteins are synthesized just-in-time, that is, not earlier than needed. Dudin et al. [14] embedded the flagellar hierarchy in the broader context of global regulatory networks in *E. coli*, notably by showing that two well-known transcription regulators, CpxR and CsgD, control the expression of flagellar genes.

In a study carried out by our group, we have focused on a central motif in this transcriptional hierarchy, consisting of the FliA and FlgM transcription factors and their targets (Fig. 5). FliA or $\sigma^{28}$ is a sigma factor that directs

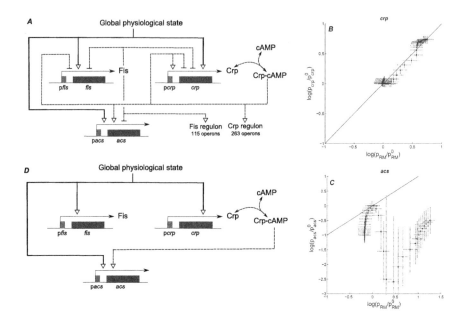

**Fig. 4. Shared control of gene expression in bacteria by transcription factors and global physiology of the cell** (adapted from [6]). *A:* Central regulatory circuit involved in the control of *E. coli* carbon metabolism, consisting of the two pleiotropic transcription factors Crp and Fis, the signaling metabolite cAMP, a target gene *acs*, and their mutual regulatory interactions. The global physiological state affects the expression of all genes in the network. *B:* Predicted and observed control of *crp* promoter activity by global physiological effetcs. Predicted (–, black) and measured (•, blue) relative activity of the *crp* promoter ($\log(p_{crp}(t)/p_{crp}^0)$) as a function of the relative activity of the $p_{RM}$ promoter ($\log(p_{RM}(t)/p_{RM}^0)$). *C:* Idem for *acs*. *D:* Reduced regulatory network, including the interactions that were found to dominate the transcriptional response of the network in *A*: the activation of all genes by the physiological state of the cell and the activation of *acs* by Crp·cAMP (Color figure online).

RNA polymerase to operons coding for the flagellar filament and the chemotaxis sensing system controlling the flagellar motor. The effect of FliA is counteracted by the anti-sigma factor FlgM. As a typical example of a FliA-dependent gene we study *tar*, which encodes a chemoreceptor protein Tar that responds to a decrease of the aspartate concentration in the medium. This signal is transmitted to downstream regulators of the flagellar motor [9]. The stability of FliA and FlgM are actively regulated, forming a check-point in the transcriptional hierarchy [4].

An often used, implicit assumption for inferring regulatory interactions from gene expression data is that it is sufficient to focus on the transcriptional level, that is, on measurements of mRNA concentrations or promoter activities. However, this brushes aside the fact that the active regulator of a gene is the protein translated from the mRNA. While the concentration of the two may be correlated to some extent at steady state [30, 45], this is certainly not the case

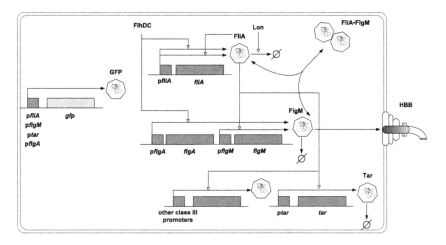

**Fig. 5. FliA-FlgM module controlling the expression of motility genes in *E. coli*** (adapted from [43]). The regulatory circuit composed of the flagellar-specific transcription factor FliA, a sigma factor also known as $\sigma^{28}$, and the anti-sigma factor FlgM forms a check-point in the transcriptional hierarchy of the motility genes in *E. coli*. FliA binds to RNA polymerase core enzyme and directs transcription from a large number of promoters [23], including p*tar* and p*flgM*. When bound to FlgM, FliA cannot activate transcription. When the hook basal-body (HBB) structure is in place, however, FlgM is exported from the cell, thus releasing FliA from the inactive complex. FliA is subject to proteolysis by Lon, but FlgM-binding protects FliA from degradation. The *fliA* promoter is auto-regulated by FliA and by a number of other regulators, most importantly the motility master regulator FlhDC. The activity of the genes in the figure is measured by fusion of their promoters to a *gfp* reporter gene on a low-copy plasmid. Genes are shown in grey or green and their promoter regions in red. Regulatory interactions are represented by open arrows, association and dissociation of FliA and FlgM as well as degradation and export by filled arrows (Color figure online).

in time-course experiments, where the time-varying concentrations of mRNA and translated protein are expected to diverge, due to the different half-lives of mRNA and protein. Typically, the half-live of mRNA is on the order of a few minutes, whereas the half-live of most proteins is longer than 10 h. We therefore asked the question to which extent the use of data on the mRNA level, instead of the protein level, biases the inference process and how this bias could be mitigated, either by adding experimental controls or by strengthening the data analysis procedures. The FliA-FlgM module provides a good test case for this study, since the regulatory interactions have been well-studied and protein turnover, an effect that is not visible on the transcriptional level, plays an important role in its functioning.

In order to address the above question, we measured the time-varying transcription of *fliA*, *flgM*, and *tar* by means of fluorescent reporter systems, consisting of transcriptional fusions of a *gfp*mut2 reporter gene to the promoters of the target genes, carried on a low-copy number plasmid [50]. Five different time-series data sets were generated, involving different strains and growth media. Specifically, we asked the question if the known model of the *tar* promoter, the expression of which is controlled by FliA and thus FlgM, can be fitted to the data when measured promoter activities are used in the identification process, instead of protein concentrations. The results are shown in Fig. 6*A* and even visual inspection shows that

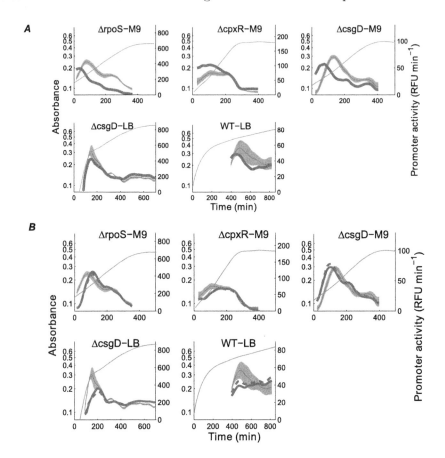

**Fig. 6. Promoter activity of** *tar* **fitted to fluorescent reporter gene data** (figure adapted from [43]). Different models of the activity of the *tar* promoter, as a function of the concentrations of FliA and FlgM and (possibly) global physiological effects, were fitted against the measured promoter activity of *tar*. The measurements were carried out in the following experimental conditions: $\Delta rpoS$ strain grown in M9 ($\Delta rpoS$-M9), $\Delta cpxR$ strain grown in M9 ($\Delta cpxR$-M9), $\Delta csgD$ strain grown in M9 ($\Delta csgD$-M9), $\Delta csgD$ strain grown in LB ($\Delta csgD$-LB), and wild-type strain grown in LB (WT-LB). *A:* Fit of a model of *tar* promoter activity when replacing the concentrations of FliA and FlgM by the measured promoter activities of *fliA* and *flgM*, respectively. Model predictions are in dark blue (thick solid line), *tar* reporter gene data are in light blue (thin solid line and shaded area). *B:* Idem, but when using protein concentrations computed from the measurement model in Sect. 3 and quantification of global physiological effects by means of a constitutive promoter.

the fit is at best of modest quality. While the fit is reasonable in two conditions, it is not satisfactory for the three others [43].

In order to test if the bad fit originates in a bias introduced by ignoring the difference in temporal expression profiles of promoter activities and protein concentrations, we developed computational procedures for reconstructing protein concentrations from promoter activities, given the (approximately) known half-lives of FliA and FlgM in the experimental conditions. These procedures build upon the measurement model of Eq. 4, which allows the promoter activity to be computed from reporter gene data and, through forward integration for a given protein half-life, the protein concentration to be predicted. In addition, following the results of the study in the previous section, we integrated global physiological effects by quantifying the activity of a constitutive promoter. The extended model, fitted against the same data, led to much improved results, shown in Fig. 6B. Additional tests showed that both extensions, reconstruction of protein concentrations and integration of global physiological effects, were necessary for obtaining the improved fits.

The results of this study suggest that, in order to reconstruct quantitative models of regulatory networks from time-series gene expression data, it is critical not to neglect the distinction between mRNA and protein. The different time-scales on which the mRNA and protein concentrations evolve causes their temporal profiles to decorrelate and makes mRNA unsuitable as a proxy for transcription factors and other regulatory proteins. While this effect has been illustrated for a single, well-studied module of the motility network, additional simulation studies have shown that the conclusions hold more generally. Due to the fact that the effective half-lives of FliA and FlgM are rather short, in comparison with most of the bacterial proteome, the consequences of using promoter activities rather than protein concentrations in the inference process are actually milder in the case of the FliA-FlgM module than in other situations [43]. While it currently remains difficult to quantify protein concentrations directly, we provided an easy-to-apply procedure for reconstructing protein concentrations from promoter activities, using the measurement models discussed in Sect. 3.1.

From a biological point of view, the results demonstrate the important role played by the active regulation of FliA and FlgM half-lives, through proteolysis and export from the cell, in shaping the dynamics of FliA-dependent promoters.

## 5   Conclusions

The above examples clearly show that reporter gene data can be used for revealing new connections in gene regulatory networks and for characterizing the dynamical adaptation of bacteria to a changing environment. In order to exploit the tool of dynamical measurements using reporter genes we need to combine an experimental strategy tailored to the specific scientific question with a generic, but flexible measurement model in order to extract the biologically relevant quantities from the data. Here, we have shown that such a measurement model can be developed and implemented in a computer tool to reliably estimate promoter activities and even protein concentrations from the primary data.

The model takes into account inherent difficulties of using fluorescent reporter genes, such as the maturation time of the fluorophore and the background fluorescence of the cell.

We have applied these tools to quantify the contribution of the global physiological state of the cell to changes in gene expression during growth transitions. Surprisingly, the response of individual genes in the network of Fig. 4 is dominated by global regulatory effects instead of specific transcription factors. This result spurs new interest in the study of the dynamics of such global phenomena. Reporter gene experiments can again provide valuable new information that would be difficult to obtain by any other technique because the phenomena we want to study are inherently dynamical. Future experiments will probably focus more on translational fusions of reporter genes to key components of the cell. This strategy will provide quantitative and dynamical measures of the "global physiological state" of the cell. The cellular components to be tagged include RNA polymerase, ribosomes, components of the DNA replication machinery, possibly chaperones, and many more. As mentioned in Sect. 2, translational fusions are more difficult to construct and generally yield signals of lower intensity. However, new and improved fluorescent reporter genes can be expected to emerge to meet these challenges.

A second application of the reporter genes shows the limits of using promoter activity, which corresponds to the concentration of mRNA, as a proxy for the activity of regulatory proteins. We show that the measurement models can be used and adapted to overcome this difficulty. Modifying the experiment may further improve the results. For example, replacing the transcriptional fusion with translational fusions will give us a more direct access to protein concentration. One way to avoid complications of long protein half-lives is to limit experiments to exponential growth where the degradation rate is dominated by growth dilution. However, such a requirement would limit the kinds of biological phenomena that can be studied.

Protein half-lives are an important parameter even for reporter genes. The measurement model developed above is independent of the absolute value of the half-life of the reporter gene. However, since promoter activity is closely related to the derivative of the reporter gene concentration, the accuracy of estimation of this variable may depend on the value of the degradation constant of the reporter gene. This parameter can be adjusted by adding a degradation tag to the end of the protein. Shorter-lived reporter genes will allow the dynamics to be estimated with greater precision, however at the cost of lower signal intensity. Moreover, the concentration of proteases involved in degradation of the reporter may be growth-rate dependent.

Another future improvement of the use of reporter genes concerns the experimental conditions for measuring the dynamics of their expression. The typical experiment involves growth transitions involving the depletion of a nutrient source. This type of experiment is easily carried out in an automated microplate reader as described here. However, in many applications we would like to change environmental conditions in a better controlled and more varied ways. For this purpose, we will need to develop online measurements of fluorescence and absorbance signals of a bacterial culture grown in a continuous fashion, for example in a chemostat.

Reporter gene measurements, combined with the appropriate measurement models and computational methods, are probably the best tool for studying and understanding the dynamics of gene regulatory networks. Future developments will further enhance the power of this tool.

**Acknowledgements.** This work was supported by the Investissements d'avenir Bioinformatique programme under project Reset (ANR-11-BINF-0005).

# References

1. Aïchaoui, L., Jules, M., Le Chat, L., Aymerich, S., Fromion, V., Goelzer, A.: BasyLiCA: a tool for automatic processing of a Bacterial Live Cell Array. Bioinf. **28**(20), 2705–2706 (2012)
2. Alon, U.: An Introduction to Systems Biology: Design Principles of Biological Circuits. Chapman & Hall/CRC, Boca Raton (2007)
3. Bansal, L., Chu, Y., Laird, C., Hahn, J.: Determining transcription factor profiles from fluorescent reporter systems involving regularization of inverse problems. In: Proceedings of the 2012 American Control Conference (ACC 2012), pp. 2725–30 (2012)
4. Barembruch, C., Hengge, R.: Cellular levels and activity of the flagellar sigma factor FliA of Escherichia coli are controlled by FlgM-modulated proteolysis. Mol. Microbiol. **65**(1), 76–89 (2007)
5. Bernstein, J., Khodursky, A., Lin, P.H., Lin-Chao, S., Cohen, S.: Global analysis of mRNA decay and abundance in Escherichia coli at single-gene resolution using two-color fluorescent DNA microarrays. Proc. Natl. Acad. Sci. USA **99**(15), 9697–9702 (2002)
6. Berthoumieux, S., de Jong, H., Baptist, G., Pinel, C., Ranquet, C., Ropers, D., Geiselmann, J.: Shared control of gene expression in bacteria by transcription factors and global physiology of the cell. Mol. Syst. Biol. **9**(634), 634 (2013)
7. Boyer, F., Besson, B., Baptist, G., Izard, J., Pinel, C., Ropers, D., Geiselmann, J., de Jong, H.: WellReader: a MATLAB program for the analysis of fluorescence and luminescence reporter gene data. Bioinf. **26**(9), 1262–1263 (2010)
8. Browning, D.F., Busby, S.J.W.: The regulation of bacterial transcription initiation. Nat. Rev. Microbiol. **2**(1), 57–65 (2004)
9. Chevance, F., Hughes, K.: Coordinating assembly of a bacterial macromolecular machine. Nat. Rev. Microbiol. **6**, 455–465 (2008)
10. Chudakov, D., Matz, M., Lukyanov, S., Lukyanov, K.: Fluorescent proteins and their applications in imaging living cells and tissues. Physiol. Rev. **90**(3), 1103–1163 (2010)
11. Cox, J., Mann, M.: Quantitative, high-resolution proteomics for data-driven systems biology. Curr. Opin. Biotechnol. **80**, 273–299 (2011)
12. de Jong, H., Ranquet, C., Ropers, D., Pinel, C., Geiselmann, J.: Experimental and computational validation of models of fluorescent and luminescent reporter genes in bacteria. BMC Syst. Biol. **4**(1), 55 (2010)

13. Dharmadi, Y., Gonzalez, R.: DNA microarrays: experimental issues, data analysis, and application to bacterial systems. Biotechnol. Prog. **20**(5), 1309–1324 (2004)

14. Dudin, O., Geiselmann, J., Oqasawara, H., Ishihama, A., Lacour, S.: Repression of flagellar genes in exponential phase by CsgD and CpxR, two crucial modulators of Escherichia coli biofilm formation. J. Bacteriol. **196**(3), 707–715 (2014)

15. Finkenstädt, B., Heron, E., Komorowski, M., Edwards, K., Tang, S., Harper, C., Davis, J., White, M., Millar, A., Rand, D.: Reconstruction of transcriptional dynamics from gene reporter data using differential equations. Bioinf. **24**(24), 2901–2907 (2008)

16. Gerosa, L., Kochanowski, K., Heinemann, M., Sauer, U.: Dissecting specific and global transcriptional regulation of bacterial gene expression. Mol. Syst. Biol. **9**, 658 (2013)

17. Giepmans, B., Adams, S., Ellisman, M., Tsien, R.: The fluorescent toolbox for assessing protein location and function. Sci. **312**(5771), 217–224 (2006)

18. Hebisch, E., Knebel, J., Landsberg, J., Frey, E., Leisner, M.: High variation of fluorescence protein maturation times in closely related Escherichia coli strains. PLoS ONE **8**(10), e75991 (2013)

19. Huang, Z., Senocak, F., Jayaraman, A., Hahn, J.: Integrated modeling and experimental approach for determining transcription factor profiles from fluorescent reporter data. BMC Syst. Biol. **2**, 64 (2008)

20. Ingolia, N.: Ribosome profiling: new views of translation, from single codons to genome scale. Nat. Rev. Genet. **15**(3), 205–213 (2014)

21. Kalir, S., McClure, J., Pabbaraju, K., Southward, C., Ronen, M., et al.: Ordering genes in a flagella pathway by analysis of expression kinetics from living bacteria. Sci. **292**(5524), 2080–2083 (2001)

22. Keren, L., Zackay, O., Lotan-Pompan, M., Barenholz, U., Dekel, E., et al.: Promoters maintain their relative activity levels under different growth conditions. Mol. Syst. Biol. **9**, 701 (2013)

23. Keseler, I., Collado-Vides, J., Santos-Zavaleta, A., Peralta-Gi, M., Gama-Castro, S., et al.: EcoCyc: a comprehensive database of Escherichia coli biology. Nucleic Acids Res. **39**, D583–D590 (2011)

24. Klumpp, S., Zhang, Z., Hwa, T.: Growth rate-dependent global effects on gene expression in bacteria. Cell **139**(7), 1366–1375 (2009)

25. Koga, K., Harada, T., Shimizu, H., Tanaka, K.: Bacterial luciferase activity and the intracellular redox pool in Escherichia coli. Mol. Genet. Genom. **274**(2), 180–188 (2005)

26. Leveau, J., Lindow, S.: Predictive and interpretive simulation of green fluorescent protein expression in reporter bacteria. J. Bacteriol. **183**(23), 6752–6762 (2001)

27. Liang, S., Bipatnath, M., Xu, Y., Chen, S., Dennis, P., Ehrenberg, M., Bremer, H.: Activities of constitutive promoters in Escherichia coli. J. Mol. Biol. **292**(1), 19–37 (1999)

28. Lichten, C., White, R., Clark, I., Swain, P.: Unmixing of fluorescence spectra to resolve quantitative time-series measurements of gene expression in plate readers. BMC Biotechnol. **14**, 11 (2014)

29. Longo, D., Hasty, J.: Dynamics of single-cell gene expression. Mol. Syst. Biol. **2**, 64 (2006)

30. Lu, P., Vogel, C., Wang, R., Yao, X., Marcotte, E.: Absolute protein expression profiling estimates the relative contributions of transcriptional and translational regulation. Nat. Biotechnol. **25**(1), 117–124 (2007)

31. Macnab, R.: Flagella and motility. In: Neidhardt, F., Curtiss III, R., Ingraham, J., Lin, E., Low, K., Magasanik, B., Reznikoff, W., Riley, M., Schaechter, M., Umbarger, H. (eds.) Escherichia coli and Salmonella: Cellular and Molecular Biology, pp. 123–45. ASM Press, Washington, DC, 2nd edn. (1996)

32. Mäder, U., Nicolas, P., Richard, H., Bessières, P., Aymerich, S.: Comprehensive identification and quantification of microbial transcriptomes by genome-wide unbiased methods. Curr. Opin. Biotechnol. **22**(1), 32–41 (2011)

33. Manen, D., Pougeon, M., Damay, P., Geiselmann, J.: A sensitive reporter gene system using bacterial luciferase based on a series of plasmid cloning vectors compatible with derivatives of pBR322. Gene **186**(2), 197–200 (1997)

34. Van Dyk, T., Wei, Y., Hanafey, M., Dolan, M., Reeve, M., Rafalski, J., Rothman-Denes, L., LaRossa, R.: A genomic approach to gene fusion technology. Proc. Natl. Acad. Sci. USA **98**(5), 2555–2560 (2001)

35. Meighen, E.A.: Bacterial bioluminescence: organization, regulation, and application of the lux genes. FASEB J. **7**(11), 1016–1022 (1993)

36. Muzzey, D., van Oudenaarden, A.: Quantitative time-lapse fluorescence microscopy in single cells. Annu. Rev. Cell. Dev. Biol. **25**, 301–327 (2009)

37. Porreca, R., Cinquemani, E., Lygeros, J., Ferrari-Trecate, G.: Structural identification of unate-like genetic network models from time-lapse protein concentration measurements. In: Proceedings of 49th IEEE Conference on Decision and Control (CDC 2010), pp. 2529–2534 (2010)

38. Ronen, M., Rosenberg, R., Shraiman, B., Alon, U.: Assigning numbers to the arrows: Parameterizing a gene regulation network by using accurate expression kinetics. Proc. Natl. Acad. Sci. USA **99**(16), 10555–10560 (2002)

39. Rowe, L., Dikici, E., Daunert, S.: Engineering bioluminescent proteins: Expanding their analytical potential. Anal. Chem. **81**(21), 8662–8668 (2009)

40. Schaechter, M., Ingraham, J., Neidhardt, F.: Microbe. ASM Press, Washington DC (2006)

41. Sharan, S.K., Thomason, L.C., Kuznetsov, S.G., Court, D.L.: Recombineering: a homologous recombination-based method of genetic engineering. Nat. Protoc. **4**(2), 206–223 (2009)

42. Silhavy, T.: Gene fusions. J. Bacteriol. **182**(21), 5935–5938 (2000)

43. Stefan, D., Pinel, C., Pinhal, S., Cinquemani, E., Geiselmann, J., de Jong, H.: Inference of quantitative models of bacterial promoters from time-series reporter gene data. PLoS Comput. Biol. **11**(1), e1004028 (2015)

44. Süel, G.: Use of fluorescence microscopy to analyze genetic circuit dynamics. Methods Enzymol. **497**, 275–293 (2011)

45. Taniguchi, Y., Choi, P., Li, G.W., Chen, H., Babu, M., Hearn, J., Emili, A., Xie, X.: Quantifying E. coli proteome and transcriptome with single-molecule sensitivity in single cells. Sci. **329**(5991), 533–539 (2010)

46. Pédelacq, J.D., Cabantous, S., Tran, T., Terwilliger, T.C., Waldo, G.S.: Engineering and characterization of a superfolder green fluorescent protein. Nat. Biotechnol. **24**(1), 79–88 (2006)

47. Tsien, R.Y.: The green fluorescent protein. Annu. Rev. Biochem. **67**, 509–544 (1998)

48. Wang, X., Errede, B., Elston, T.: Mathematical analysis and quantification of fluorescent proteins as transcriptional reporters. Biophys. J. **94**(6), 2017–2026 (2008)

49. Young, J., Locke, J., Altinok, A., Rosenfeld, N., Bacarian, T., Swain, P., Mjolsness, E., Elowitz, M.: Measuring single-cell gene expression dynamics in bacteria using fluorescence time-lapse microscopy. Nat. Protoc. **7**(1), 80–88 (2011)

50. Zaslaver, A., Bren, A., Ronen, M., Itzkovitz, S., Kikoin, I., Shavit, S., Liebermeister, W., Surette, M., Alon, U.: A comprehensive library of fluorescent transcriptional reporters for Escherichia coli. Nat. Methods **3**(8), 623–628 (2006)

51. Zulkower, V., Page, M., Ropers, D., Geiselmann, J., de Jong, H.: Robust reconstruction of gene expression profiles from reporter gene data using linear inversion. Bioinf. **31**(12), i71–i79 (2015)

# Modeling and Analysis of Qualitative Behavior of Gene Regulatory Networks

Alvis Brazma[1], Karlis Cerans[2], Dace Ruklisa[3], Thomas Schlitt[4], and Juris Viksna[2(✉)]

[1] European Molecular Biology Laboratory, European Bioinformatics Institute,
EMBL-EBI, Hinxton, UK
brazma@ebi.ac.uk
[2] Institute of Mathematics and Computer Science, Riga, Latvia
{karlis.cerans,juris.viksna}@lumii.lv
[3] University of Cambridge, Cambridge, UK
dr320@medschl.cam.ac.uk
[4] King's College London, London, UK
thomas@thomas-schlitt.net

**Abstract.** We describe a hybrid system based framework for modeling gene regulation and other biomolecular networks and a method for analysis of the dynamic behavior of such models. A particular feature of the proposed framework is the focus on qualitative experimentally testable properties of the system. With this goal in mind we introduce the notion of the frame of a hybrid system, which allows for the discretisation of the state space of the network. We propose two different methods for the analysis of this state space. The result of the analysis is a set of attractors that characterize the underlying biological system.

Whilst in the general case the problem of finding attractors in the state space is algorithmically undecidable, we demonstrate that our methods work for comparatively complex gene regulatory network model of $\lambda$-phage. For this model we are able to identify attractors corresponding to two known biological behaviors of $\lambda$-phage: *lysis* and *lysogeny* and also to show that there are no other stable behavior regions for this model.

## 1 Introduction

Hybrid systems (HS) are a natural choice for modeling biomolecular networks for at least two reasons: (1) they can model processes that are relevant to behavior of biomolecular networks – they can describe both discrete aspects (e.g. states of activity of specific promoters) and continuous aspects (e.g. concentrations of biological substances in a cell); (2) well established mathematical techniques and supporting software tools exist for analysis of such hybrid system models. One of the first explicit applications of a HS based approach to the modeling of biomolecular networks has been described by Alur et al. in [2], where the authors discuss

The authors are listed in alphabetical order and have equally contributed to the paper. The work was supported by Latvian Council of Science grant 258/2012 and Latvian State Research programme project NexIT (2014-2017).

© Springer International Publishing Switzerland 2015
O. Maler et al. (Eds.): HSB 2013 and 2014, LNBI 7699, pp. 51–66, 2015.
DOI: 10.1007/978-3-319-27656-4_3

a rather general class of HS models and show that such models are adequate for description and simulation of biological networks.

There is a significant number of other studies discussing applications of HS to biomolecular network modeling, often proposing somewhat more restricted formalisms than the one used in [2] and providing examples of applications of these models to description of specific biological systems (see for example [1, 4, 7, 9, 15], which is by no means a comprehensive list). One of the most recent of such studies [8] describes an HS based Temporal Evolution Model and applies it to modeling of *Drosophila* circadian cycle. Multiaffine Hybrid Automata models ([3, 10]) that correspondingly have been applied to cardiac cell and bone cell modeling technically are similar to our approach, however the emphasis is on the simulation and identification of parameter values.

Whilst not stated in terms of HS explicitly, a related approach has been presented in [16, 17]. These models describe a biological system using differential equations and then analyze stability of specific cyclic behaviors ('circuits') at a logical level. Notably, by using this approach the stability of several regulatory circuits for $\lambda$-phage has been shown ([17]).

Our work presented here is generally in line with previous studies of application of hybrid systems to biomolecular network modeling and is motivated by two observations. First, it can be experimentally difficult to measure the quantitative parameters of biological systems accurately and experimental results often are closer to a qualitative assay than a quantitative measurement (e.g., it may be possible to detect if the concentration of a particular substance is increasing or decreasing, while measuring the exact rate is much more difficult). Second, in some cases it is possible to separate the structure of the underlying regulatory network from its quantitative parameters. In this case it is natural to ask to what extent the qualitative behavior of the system depends on the structure of the network alone, and to what extent on the exact quantitative values (relative or absolute) of the parameters.

Driven by these assumptions we propose a Hybrid System Model (HSM) tailored to the description of biomolecular networks and gene regulatory networks in particular. HSM can be viewed as a restricted version of hybrid system that still provides sufficient power for modeling of biological systems, while the restrictions imposed upon HSM facilitate the analysis of the models. HSM is a generalization of the authors' previous work on Finite State Linear Model (FSLM) ([5, 12, 13]).

A variation of HSM model and its application to the analysis of behavior of gene network of $\lambda$-phage ([11]) has been previously described by the authors in [6]. Here we present a more developed mathematical formalism for separation between quantitative and qualitative parameters of the system: (1) we assume that a biological system is correctly represented by a HSM, however the parameter values are unknown; (2) known however is a structure (modes and transitions between them) of HSM represented by its *frame*; (3) the analysis of HSM behavior is done at the level of *constrained frames* in which the exact parameter values are replaced by discrete constraints on them.

We also present a new algorithm for analysis of the *universal state space* of the constrained frames of HSM. This allows us to derive the constraints that affect the behavior of the system in single process of universal state space analysis, avoiding exhaustive analysis of state spaces for all the possible sets of constraints that has been done previously. The mathematical formalism is presented here also in mathematically more rigorous terms than in the previous work.

For the assessment of the merits of the modeling and analysis methods described here, we have applied the model to a well-studied gene network of $\lambda$-phage [11]. $\lambda$-phage is a bacterial virus, which when invading its host can exhibit two different stable behaviors *lysis* and *lysogeny.*

It should be noted that due to the undecidability of the reachability problem for HSM we can not guarantee that HSM state space analysis will provide any results. Therefore it is noteworthy that for a comparatively complicated gene regulatory network model of $\lambda$-phage our method was able to identify two regions of stable behavior in the model's state space that correspond to the two biologically known behaviors – lysis and lysogeny. Moreover, these are the only regions of stable behavior and their existence does not depend on the exact quantitative parameters, but only on the structure of the network and the experimentally known qualitative information.

## 2  Hybrid Systems for Modeling Gene Regulatory Networks

Like in most previous approaches that use HS for modeling gene regulatory networks (GRN), we use *modes* to represent different combinations of transcription factor binding site states (a binding site may be either vacant or in an occupied state) and continuous variables for representing the concentrations of various biological substances (e.g. proteins) in a cell. We assume that in each mode substance concentrations change according to continuous functions from a given set of possible functions. The change of mode is defined by a condition in transition diagram; mode is changed when a concentration of a substance reaches a certain threshold, i.e. one of the predicates 'guarding' a transition from the given mode is satisfied. In gene regulatory networks these thresholds correspond to association or dissociation concentrations of proteins that have to be reached in order to bind or dissociate from a particular binding site.

We keep our HSM formalism as simple as possible, but still sufficient for modelling biological processes. Most notably we disallow instantaneous resetting of continuous values to 0, since such resetting does not seem to have any valid biological interpretation.

**Definition 1.** *A  Hybrid  System  Model  (HSM)  is  a  6-tuple  $\mathcal{H}$  =*  $\langle M, X, C, T, F, MF \rangle$, *where:*

1. $M = \{\mu_1, \ldots, \mu_k\}$ *is a finite set of* modes.
2. $X = \{x_1, \ldots, x_m\}$ *is a finite set of* continuous variables *that can assume real non-negative values.*
3. $C = \{c_1, \ldots, c_r\}$ *is a finite set of real non-negative* transition constants.

4. *T is a set of* mode transitions, *where each transition* $\tau \in T$ *has the form* $\tau = \alpha \to_p \beta$, *where* $\alpha, \beta \in M$ *and* $p = p(\tau)$ *is a predicate that has a form* $x \leq c$ *or* $x \geq c$ *for some variable* $x = x(\tau) \in X$ *and some constant* $c = c(\tau) \in C$. *Predicate* $p$, *called a* guard, *is a function* $p : \mathbf{R}_+ \to \{true, false\}$ *the value of which depends on the value of variable* $x$.
5. $F = \{f_1, \ldots, f_n\}$ *is a set of real non-negative two argument* growth/degradation *functions* $f_i : \mathbf{R}_+ \times \mathbf{R}_+ \to \mathbf{R}_+$ *that are continuous and monotonous in both arguments and for which* $f_i(z, 0) = z$ *for every* $z \in \mathbf{R}_+$.
6. $MF : M \times X \to F$ *is a mapping providing* mode-function assignments *assigning to each mode* $\alpha \in M$ *and each variable* $x \in X$ *a function* $g \in F$.

A 'toy example' of hybrid system model of a GRN with two genes and three protein binding sites is shown in Fig. 1.

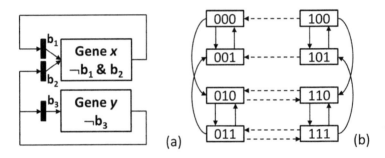

**Fig. 1.** A 'toy example' of GRN with two genes and three protein binding sites (a), and the corresponding HSM with 8 modes (b). Each binding site $b$ becomes occupied if the concentration of gene product binding to it reaches an association constant $a_b$ and becomes vacant if the concentration drops to dissociation constant $d_b$ (where $a_b > d_b$). HSM modes are denoted by binding site states (e.g. 100 represents the situation when site $b_1$ is occupied and $b_2$ and $b_3$ are vacant); no concrete growth functions are shown, but functions are increasing if the value of a Boolean function from the states of binding sites is *true* and decreasing if *false*. Transitions triggered by changes of variable $x$ are shown with dashed and transitions triggered by changes of variable $y$ with solid lines, constants in transition guards are not shown.

Intuitively a mode-function assignment assigns to each mode $\alpha \in M$ and each variable $x \in X$ a growth/degradation function that describes the change of this variable in time. Whilst these functions could be viewed as a part of the corresponding modes, if we are concentrating on qualitative aspects of such functions (e.g. are they increasing or decreasing) and not on their precise form, it may be useful to distinguish between pairs of modes with identical functional assignments for a given variable and pairs of modes with different ones.

Each growth/degradation function $f$ has two real valued arguments. The first of them corresponds to the value of a variable $x$ at the time of switching to a particular mode for which $f$ is assigned to be the function regulating $x$, while the second is the time $\Delta t$ elapsed since the switch. For instance, in a special case when the growth rates are constant, the functions are linear and have a form $g(x, \Delta t) = x + c \cdot \Delta t$, where $c$ is a constant. For these particular functions the condition $g(z, 0) = z$ is also satisfied. This condition ensures the continuity of trajectories when the system switches from one mode to another.

In comparison with more general definitions of hybrid systems we have placed several restrictions on the set of functions according to which the substance concentrations can change. One such restriction is that only the mode of the system and the value of a particular substance determines the rate of concentration change until a switch to a different mode happens. In addition in a given mode the concentration of a particular substance can only increase, decrease or stay constant. In practice we also need to impose a few additional restrictions (mostly too technical to merit inclusion in HSM definition) to exclude some 'undesirable behaviors' of the system. Most notably we require that if concentration of some substance is moving towards a transition constant triggering a guard then eventually this constant will be reached. The formalism is strong enough to describe biological systems, but does not provide more freedom for the behavior of the system than is necessary.

To describe the behavior of a HSM starting at a given initial mode with a given set of initial variable values we use a concept of *run*. A run describes an evolution of HSM over time by specifying a concrete sequence of modes through which the system is evolving and assigning well-defined values $x(t)$ to each of the variables $x \in X$ at time moments $t$.

Let $\mathcal{H} = \langle M, X, C, T, F, MF \rangle$ be a HSM with given initial mode $\alpha_0$, initial time moment $t_0$ and initial values of variables $X(t_0) = (x_1(t_0), \ldots, x_m(t_0))$. We define a *run* of the system $\mathcal{H}$ as a (finite or infinite) sequence of modes $\alpha_i \in M$ and times $t_i$: $\mathcal{R}(\alpha_0, t_0, X(t_0)) = (\alpha_0, t_0) \rightarrow (\alpha_1, t_1) \rightarrow (\alpha_2, t_2) \rightarrow \cdots$.

For each $(\alpha_i, t_i)$ $t_i$ is the time point when $\mathcal{H}$ switches to the mode $\alpha_i$. While the system is in the mode $\alpha_i$ its variables change as defined by the mode function assignment in that mode (i.e. for all $j$: $x_j(t) = g_j(x_j(t_i), t - t_i)$, where $g_j = MF(\alpha_i, x_j)$ – a function assigned to $x_j$ in mode $\alpha_i$). Such an evolution continues until one of the guards is satisfied (if several are satisfied simultaneously, we can assume that one is selected by some deterministic procedure).

A particular run $\mathcal{R}(\alpha_0, t_0, X(t_0))$ assigns well-defined functions describing changes of the variables in time. For each $x_j \in X$ and $t \in [t_i, t_{i+1}]$ it defines $x_j(t) = g_j(x_j(t_i), t - t_i)$ and thus also defines a vector function $X = (x_1, \ldots, x_m)$ in the whole interval $[t_0, \infty[$.

Thus a specific run of $\mathcal{H}$ describes a precise and (in principle) experimentally measurable behavior of the system, and given an appropriate HSM for some biological system, a run can be regarded as simulation of the behavior of this system starting from some known initial conditions. Still, if we want to describe all the possible behaviors of the system in such a way, we (normally) need a continuum of different runs.

A natural and well explored alternative is to disregard the exact values of function $X$ describing changes of variables, but consider only sequences of modes $\alpha_0, \alpha_1, \ldots$ that can occur in runs. Provided that modes of HSM have well defined biological interpretation, modes can also be more easily determined in experiments than exact concentrations of substances. Still, even experimental measurement of modes may not be a simple task.

We define a *path* of HSM in order to describe a finite sequence of modes that can occur in a particular run: a finite sequence $\alpha_0, \ldots, \alpha_n$ is called a *path* if there is a run $\mathcal{R}(\alpha_0, t_0, X(t_0)) = (\alpha_0, t_0) \rightarrow \cdots \rightarrow (\alpha_n, t_n) \rightarrow \cdots$.

# 3   Qualitative Behavior of Networks and Frames of HSM

An appropriate HSM $\mathcal{H}$ can provide a good approximation of a biological system. However such a model also involves a large number of quantitative parameters: the set of growth/degradation functions and the set of the guards governing transitions, the knowledge of which is only rarely a realistic assumption.

Nevertheless usually we can assume that a set of modes $M$ is known – a separate mode can be assigned to each state of gene activity (active or not) and/or binding site state (occupied or not). In addition it is often possible to define the set of variables $X$ (e.g. the set of substances in whose concentrations we are interested), and gather the information about the transitions in $T$ (the modes and variables involved) and a partial information about the growth and degradation functions (e.g. whether a particular concentration is growing or decreasing). Guards in our model have a very simple form: either $x \leq c$ or $x \geq c$. Often we know the type of the inequality, but not the exact constant $c$.

To specify a HSM using such a limited information (more of qualitative than quantitative nature) about the system we introduce the notion of *frame*. A single frame is intended to represent a whole set of HSMs that are consistent with the existing knowledge about the system.

A *frame* of HSM is defined as a 5-tuple $\mathcal{F} = \langle M, X, C', T, MF' \rangle$, where the requirements on $M, X, T$ are the same as in the definition of HSM, but we don't have a set of concrete functions $F$ and use an assignment $MF' : M \times X \rightarrow \{\nearrow, \searrow, \rightarrow\}$ instead⋆. In addition for frames a set $C'$ is not a set of constants, but instead is a set of *variables* assuming real non-negative values. For two different transitions $\tau_1$ and $\tau_2$ the notation $c(\tau_1) = c(\tau_2)$ indicates that the 'constants' involved in the guards of these transitions are the same, otherwise it is assumed that these 'constants' are distinct. Normally we will use $c(\tau_1) = c(\tau_2)$ only for transitions with $x(\tau_1) = x(\tau_2)$.

Thus essentially a frame is a simplified HSM, for which only qualitative information about the growth/degradation functions and the guards is specified.

For a given HSM $\mathcal{H}$ we can easily construct the frame by using the same sets of modes $M$, variables $X$ and transitions $T$ and simply replacing $C$ and $MF$ with $C'$ and $MF'$ providing less restricted information about the transition guards and growth functions. We call such a frame an *induced* frame of $\mathcal{H}$ and denote it by $\mathcal{F}(\mathcal{H})$.

The concept of *run* can be extended to frames. Since for frames we lack information about substance concentrations, their change rates and their relation

---

⋆ For biomolecular networks the value $"\rightarrow"$ describing the situation where concentration of some substance does not change is generally reserved for the cases in which concentration is either 0 or the maximal biologically feasible saturation value.

to transition guards, for frames each run is specified only by an initial mode $\alpha_0$ and is just a sequence of modes $\mathcal{R}(\alpha_0) = \alpha_0 \rightarrow \alpha_1 \rightarrow \alpha_2 \rightarrow \cdots$, where $\alpha_i \rightarrow \alpha_{i+1}$ is allowed if and only if there is a transition $\tau = \alpha_i \rightarrow_p \alpha_{i+1} \in T$.

Runs for frames are generally non-deterministic, i.e. not uniquely specified by initial $\alpha_0$. Usually runs will be infinite sequences of modes, unless they terminate with some mode $\beta$ that does not have outgoing transitions. Consistently with the terminology used for HSM a finite sequence $\alpha_0, \ldots, \alpha_n$ is called a *path* if there is a run $\mathcal{R}(\alpha_0)$ containing this sequence as an initial fragment.

In general frame $\mathcal{F} = \mathcal{F}(\mathcal{H})$ will have a multitude of paths $\alpha, \ldots, \beta$ in $\mathcal{F}$ for which there are no corresponding paths (i.e. with identical mode sequences) in $\mathcal{H}$. Moreover, for some paths $\alpha, \ldots, \beta$ in $\mathcal{F}$ there may not exist any HSM $\hat{\mathcal{H}}$ with $\mathcal{F} = \mathcal{F}(\hat{\mathcal{H}})$ in which $\alpha, \ldots, \beta$ is a path (the reason is that for frames we do not have the means of enforcing a consistent behavior of growth/degradation functions each time a particular mode is encountered in the run). Thus in general frame runs may describe behaviors that are not consistent with the qualitative information about the growth/degradation functions of the system included in the specification of its frame. However, it is evident that for each run $\mathcal{R}(\alpha_0, t_0, X(t_0))$ of $\mathcal{H}$ there always will be a run $\mathcal{R}(\alpha_0)$ of $\mathcal{F}(\mathcal{H})$ with the same sequence of modes (the same will hold also for paths). Therefore, we do not lose any of systems' behaviors when considering frames instead of HSM.

Whilst for a given HSM $\mathcal{H}$ its behavior can be represented by its induced frame, for analysis purposes it may be useful to characterize $\mathcal{H}$ with other frames that have more fine grained structure. To do this we introduce the notion of frame *refinement*.

**Definition 2.** *A frame* $\mathcal{F}_1 = \langle M_1, X, C', T_1, MF_1' \rangle$ *is a* refinement *of frame* $\mathcal{F}_2 = \langle M_2, X, C', T_2, MF_2' \rangle$ *(denoted* $\mathcal{F}_1 < \mathcal{F}_2$*), if there are surjective mappings* $m : M_1 \rightarrow M_2$ *and* $t : T_1 \rightarrow T_2$ *such that:*

1. *for all* $\alpha \rightarrow_p \beta \in T_1$: $t(\alpha \rightarrow_p \beta) = m(\alpha) \rightarrow_p m(\beta) \in T_2$;
2. *for all* $\alpha \in M_1$ *and all* $x \in X_1$: $MF_1'(\alpha, x) = MF_2'(m(\alpha), x)$;
3. *the sequence* $m(\alpha_0), \ldots, m(\alpha_n)$ *is a path in frame* $\mathcal{F}_2$ *for any path* $\alpha_0, \ldots, \alpha_n$ *in frame* $\mathcal{F}_1$.

Thus we allow partitioning of modes and removal of some transitions from partitioned modes in a frame refinement as long as all the paths of the original frame are preserved. Therefore by the definition of a refinement all the 'behaviors' of the initial frame will also be exhibited by its refinement.

In a similar manner we say that a frame $\mathcal{F}$ *supports* HSM $\mathcal{H}$ (denoted by $\mathcal{F} \triangleleft \mathcal{H}$), if for all paths in $\mathcal{H}$ there are corresponding paths in supporting frame $\mathcal{F}$. It is evident that frame refinement is transitive and that any HSM $\mathcal{H}$ will always be supported by the induced frame $\mathcal{F}(\mathcal{H})$.

In HSM models of biological networks guards of transitions $x \geq c$ $(x \leq c)$ can correspond to an event when the concentration of a protein described by a variable $x$ reaches an association constant of some site and binds to it (or drops below a dissociation constant and vacates the site). The exact values of binding site affinities usually are unknown, however in cases where there are several binding sites for the same protein at least a partial ordering of binding affinities may be known. To characterize such affinity orderings we use the notion of *constraints*. For a given frame $\mathcal{F} = \langle M, X, C', T, MF' \rangle$ we define a *constraint* $O(C')$ as a transitive directed acyclic graph with a set of vertices $C'$. A constraint effectively specifies a *strict* (i.e. non-reflexive) partial ordering of values in $C'$: for $c_1, c_2 \in C'$ an edge $(c_1, c_2)$ in constraint $O(C')$ is interpreted as inequality $c_1 < c_2$. Normally graphs $O(C')$ will consist of a number of non-connected components, a separate one for each variable in $X$. We denote by $Cons(\mathcal{F})$ the set of all constraints of frame (i.e. set of all partial orderings of $C'$). By $O_\emptyset(C')$ we denote constraint with no edges.

Given two constraints $O_1(C')$ and $O_2(C')$ with the same set of vertices $C'$ and sets of edges $E(O_1)$ and $E(O_2)$ the union $O_1 \cup O_2$ denotes a graph with the vertex set $C'$ and the set of edges a transitive closure of $E(O_1) \cup E(O_2)$. If $O_1 \cup O_2$ contains a cycle, it is not a constraint and we say that constraints $O_1$ and $O_2$ are *incompatible*. Otherwise we say that $O_1$ and $O_2$ are *consistent*. Similarly the intersection of constraints $O_1 \cap O_2$ denotes a graph with the set of edges $E(O_1) \cap E(O_2)$. $O_1 \cap O_2$ is always a constraint.

**Definition 3.** *A* constraint *frame* *is a pair* $(\mathcal{F}, CA)$, *where* $\mathcal{F} = \langle M, X, C', T, MF' \rangle$ *is a frame and* $CA : T \to Cons(\mathcal{F})$.

We say that two constraint assignments $CA_1$ and $CA_2$ are *incompatible* if for some transition $\tau$ constraints $CA_1(\tau)$ and $CA_2(\tau)$ are incompatible. Otherwise we say that $CA_1$ and $CA_2$ are *consistent*. For consistent $CA_1$ and $CA_2$ it is convenient to denote by $CA_1 \cup CA_2$ the assignment of constraint $CA_1(\tau) \cup CA_2(\tau)$ to each transition $\tau$. We can define $CA_1 \cap CA_2$ analogously. By $CA_O$ we denote the assignment of the same constraint $O$ to all transitions.

In constrained frame each transition is annotated with a partial ordering of $C'$ and we consider transition as 'available' only if its transition constant has the highest priority among all the transitions from the same mode and involving the same variable. Moreover we require that in constrained frames a sequence of modes forming a run should be 'passable' by a sequence of transitions that does not include any transition pairs with incompatible constraint assignments.

To make this more precise we define constraint $O(\alpha, \beta)$ as an intersection of all constraints $CA(\tau)$ for transitions $\tau = \alpha \to_p \beta \in T$ for a particular pair of modes $\alpha, \beta$ in constrained frame $(\mathcal{F}, CA)$. Furthermore, for each sequence of modes $P = \alpha_0, \ldots, \alpha_n$ we define a graph $O(P)$ as a union of all constraints $O(\alpha_i, \alpha_{i+1})$, $i = 0 \ldots n - 1$. Informally, if $O(P)$ is a constraint (i.e. without cycles) it can be regarded as the least restrictive *single* constraint under which the sequence of modes $P$ is 'passable'.

For constrained frames it is convenient to start with a definition of path instead of run – a finite sequence $P = \alpha_0, \ldots, \alpha_n$ is a *path* if $O(P)$ is a constraint. Then $\mathcal{R}(\alpha_0) = \alpha_0 \to \alpha_1 \to \alpha_2 \to \cdots$ is a *run* if each initial fragment of sequence of its modes $\alpha_0, \ldots, \alpha_n$ is a path. For finite runs $\mathcal{R}(\alpha_0) = \alpha_0 \to \cdots \to \alpha_n$ we also require that $\mathcal{R}(\alpha_0)$ is not a proper prefix of some other run $\hat{\mathcal{R}}(\alpha_0)$.

The notion of frame refinement can be extended to constrained frames. The definition of a constrained frame $(\mathcal{F}_1, CA_1)$ being a refinement of $(\mathcal{F}_2, CA_2)$ (denoted by $(\mathcal{F}_1, CA_1) < (\mathcal{F}_2, CA_2)$) is analogous to Definition 2 with an additional requirement that a mapping $t$ of transitions should be consistent with the constraint assignments (i.e. $CA_1(t(\tau)) = CA_2(\tau)$) and the requirement for path conservation now referring to paths in constrained frames.

Let us consider HSM $\mathcal{H}$ describing some gene regulatory network. The complete ordering of affinities $C$ of binding sites in $\mathcal{H}$ is represented by some constraint $O$ and it would be adequate to describe this GRN with induced frame $\mathcal{F}(\mathcal{H})$ and constraint assignment $CA_O$. We call such constraint assignment $CA_O$ *maximal* and for given $\mathcal{H}$ denote it by $Max(\mathcal{H})$. The induced constrained frame will be denoted by $(\mathcal{F}(\mathcal{H}), Max(\mathcal{H}))$. This notation allows to extend definition of frame support for HSM to constrained frames: constrained frame $(\mathcal{F}, CA)$ supports $\mathcal{H}$ (denoted by $(\mathcal{F}, CA) \triangleleft \mathcal{H}$) if $(\mathcal{F}, CA) < (\mathcal{F}(\mathcal{H}), Max(\mathcal{H}))$.

However usually we do not have complete knowledge of ordering of $C$. In the worst case when we do not possess any knowledge about the orderings of affinities we can only consider the least constrained frame that will support $\mathcal{H}$ regardless of the affinity ordering in $\mathcal{H}$. In such frame we assign to all transitions in $\mathcal{F}$ constraints that give them the highest priority – i.e. to each transition $\alpha \rightarrow_p \beta \in T$ with guard $p$ of the form $x(\tau) \le c(\tau)$ we assign a constraint with edge set $\{(c(\tau), c(\hat{\tau})) | \hat{\tau} = \alpha \rightarrow_p \gamma \in T, \gamma \in M, \tau \neq \hat{\tau}, x(\tau) = x(\hat{\tau})\}$ (or edge set $\{(c(\hat{\tau}), c(\tau)) | \hat{\tau} = \alpha \rightarrow_p \gamma \in T, \gamma \in M, \tau \neq \hat{\tau}, x(\tau) = x(\hat{\tau})\}$ if $p$ is of form $x(\tau) \ge c(\tau)$). We call the resulting constraint assignment *minimal* and for frame $\mathcal{F}$ denote it by $Min(\mathcal{F})$. In contrast to maximal assignment, which is based on underlying HSM $\mathcal{H}$, minimal assignment is defined solely by the properties of $\mathcal{F}$.

# 4   Analysis of Dynamic Behavior of Hybrid System Models

Given a constrained frame $(\mathcal{F} = \langle M, X, C', T, MF' \rangle, CA)$ describing some system, all its possible behaviors are represented by runs that are allowed under the specified constraints. We can conveniently characterize all such runs by a graph whose vertices correspond to modes of $\mathcal{F}$ and edges correspond to the transitions in $\mathcal{F}$ that are allowed in runs by a constraint assignment $CA$. We call such graph a *state space graph* of $\mathcal{F}$ and denote it by $G = G(\mathcal{F}, CA)$. The vertex set of $V(G)$ is simply $M$, the set of edges $E(G)$ is subset of $\{(\alpha, \beta) | \alpha \rightarrow_p \beta \in T\}$. There is a simple algorithm that computes $G(\mathcal{F}, CA)$ for given $\mathcal{F}$ and $CA$.

There are notable similarities between frame state space graphs and state space graphs in Boolean gene network models. For Boolean models the analysis of their space graphs is also simple and gives unambiguous characterization of the system's behavior – the graphs decompose into cyclic *attractor* subgraphs, each of which can be regarded as a descriptor of one of the possible behaviors of the system. The information that a frame state space graph provides about the system's behavior is very similar to that given by state space graphs of Boolean models. Unfortunately, however, frame state space graphs can be much more complex and do not allow for a simple partitioning into attractor basins.

We are interested in identifying the regions in frame state space graphs $G = G(\mathcal{F}, CA)$ that will characterize 'stable behaviors' of the system. In order to achieve this task we propose to partition $G$ into *strongly connected components*

(SCC) and to relate these components to stable behaviors (similar generalization of attractors has been already used for Random Boolean Networks in [14]).

In the worst case the whole $G$ can consist just of a single SCC. However we can perform a more detailed analysis of the dynamics than just compute a partition of $G$ into SCCs. Let us consider a SCC $S \subseteq V(G)$ and a variable $x$, such that for all modes $\alpha \in S$: (1) we have $MF'(\alpha, x) = \nearrow$ (or $MF'(\alpha, x) = \searrow$), and (2) there is an edge $(\alpha, \beta) \in E(G)$ derived from transition $\tau = \alpha \rightarrow_p \beta$ with the guard of the form $x \leq c(\tau)$ (correspondingly $x \geq c(\tau)$). In such a situation we can conclude that the system can stay in $S$ only for a limited time, since eventually one of guards for these transitions will get satisfied (due to restrictions we impose on growth functions). In such cases we say that SCC $S$ is *transitional*.

**Definition 4.** *A strongly connected non-transitional component of constrained state space graph $G(\mathcal{F}, CA)$ is called an* attractor.

Finding of attractors requires splitting state space graphs in SCCs and checking whether each SCC is or is not transitional. The latter task can be achieved in linear time with respect to the size of graph.

Suppose that we have HSM model $\mathcal{H}$ for some gene regulatory network. We assume that we have only qualitative knowledge about parameters of $\mathcal{H}$, i.e., we have complete knowledge whether growth/degradation functions are increasing or decreasing and possibly a partial knowledge about the ordering of transition constants. Such amount of available information seems to be typical for many biological networks. In terms of constrained frames our knowledge about the system is represented by an induced constrained frame $(\mathcal{F}(\mathcal{H}), CA)$ with a set of constraints $CA$ ranging somewhere between $Min(\mathcal{F}(\mathcal{H}))$ (no knowledge about transition constants) and $Max(\mathcal{H})$ (complete knowledge about transition constants). Usually however some information about constant ordering is available, e.g. if constants correspond to association and dissociation affinities of binding sites, then for the same binding site association affinity must be the largest of these two. Additional constraints also can be derived from the known biological facts. We denote these known constraint assignments by $External(\mathcal{H})$.

In order to make some judgments about possible $\mathcal{H}$ behaviors we are interested in finding the sets of all attractors of all frames $(\mathcal{F}(\mathcal{H}), CA)$ for which $CA$ is consistent with $External(\mathcal{H})$. There are two natural ways to do this.

Firstly, we could consider all *complete orderings $Ord \in Cons(\mathcal{F}(\mathcal{H}))$* of transition constants that are consistent with $External(\mathcal{H})$ and analyze attractor structure of all the corresponding graphs $G(\mathcal{F}(\mathcal{H}), CA_{Ord})$ (thus essentially we check for attractor structure of graphs corresponding to all the possible choices of $Max(\mathcal{H})$). Such an approach has been used by the authors in [6] for analysis of $\lambda$-phage model. Despite comparatively large size of this model (11664 modes and 32 transition constants) the number of different orderings that have to be considered is comparatively small – only 42. The main reason for this is (easily provable) fact that attractor structure is influenced only by orderings of subsets of constants $\{c(\tau)|x(\tau) = x\}$ for each variable $x \in X$. Therefore for HSM models where transitions are defined by binding site affinities, in the special case when for each binding factor there is only single binding site it affects, there will be only a single ordering $Ord$ to consider. For our $\lambda$-phage model there are multiple binding sites for most of the binding factors and 'biologically known' constraints are used to reduce the number of possible orderings from a few thousands to 42. One of the results we have presented in [6] is the fact that the attractor structure remains the same for all these 42 orderings and contains only 2 attractors that correspond to two known biological behaviors of $\lambda$-phage: *lysis* and *lysogeny*.

An alternative approach on which we focus in this paper is to try to analyze directly attractor structure $G(\mathcal{F}, CA)$ using only known limited knowledge of $CA$ that is given by $External(\mathcal{H})$ and $Min(\mathcal{F}(\mathcal{H}))$ without explicit consideration of all the possible choices of $CA$. Such an approach has several advantages: first, if two different assignments $CA_1$ and $CA_2$ have similar state space graphs we probably can save some work by noticing shared parts in these graphs; second this could help to decide whether the behaviors described by the attractors yielded by different choices of $CA$ are essentially the same or different; third such an approach could help to derive automatically the conditions (i.e. constraint assignments) that separate different qualitative behaviors.

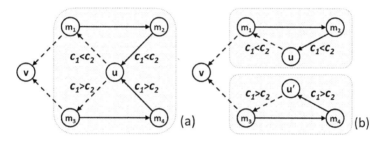

**Fig. 2.** A simple example showing that refinement of constrained frame (b) can have attractors with fewer number of states than in original frame (a). In original frame there is a SCC containing states $m_1, m_2, m_3, m_4, u$, which have incompatible constraints for two different paths between states $u$ and $v$. In refinement this SCC has been split in two SCCs with states $m_1, m_2, u$ and $m_3, m_4, u'$. Solid and dashed lines represent correspondingly transitions and paths, in the latter case the constraints shown refer to the whole paths.

To perform such analysis we can start with initially given state space graph $G_0 = G(\mathcal{F}(\mathcal{H}), Min(\mathcal{F}(\mathcal{H})) \cup External(\mathcal{H}))$, check for parts of $G_0$ that will 'behave' differently if different additional restrictions are imposed on existing constraints, and try to partition these parts in such a way that different behaviors are represented by different parts of these partitions. Let $G_1$ be a graph obtained by such a process from $G_0$. By repeating this process we will obtain (hopefully finite) sequence of graphs $G_0, G_1, \ldots, G_n$, where $G_n$ can not be further partitioned. By analyzing attractor structure of $G_n$ we can then expect to find attractors for all different choices of constraints compatible with initial constraints $Min(\mathcal{F}(\mathcal{H})) \cup External(\mathcal{H})$ and, moreover, hope, although that is not guaranteed, that some attractors will be shared by several orderings of constants.

It turns out that the process of computing space graph sequence $G_0, G_1, \ldots, G_n$ essentially can be regarded as a process of constructing appropriate constrained frame refinements $(\mathcal{F}_0, CA_0), \cdots, (\mathcal{F}_n, CA_n)$, where $(\mathcal{F}_0, CA_0) = (\mathcal{F}(\mathcal{H}), Min(\mathcal{F}(\mathcal{H})) \cup External(\mathcal{H}))$, $(\mathcal{F}_n, CA_n) < \cdots < (\mathcal{F}_0, CA_0)$ and $G_n = G(\mathcal{F}_n, CA_n)$. An example in Fig. 2 shows that such a refinement process indeed can reduce the number of states in the attractors of a state space graph. For computing the sequence of graphs we propose the *RefineStateSpace* algorithm. We use assignments $init(\alpha)$ and $constr(\alpha)$ to refer correspondingly to the initial mode from which $\alpha$ has been derived and to the last transition constraint that has triggered creation of $\alpha$.

**Algorithm .** *RefineStateSpace*
**Input:** State space graph $G(\mathcal{F}_0, CA_0)$
**Output:** Refined state space graph $G(\mathcal{F}, CA)$
1. $\mathcal{F} = \langle M, X, C', T, MF' \rangle \leftarrow \mathcal{F}_0$, $CA \leftarrow CA_0$
2. for each $\alpha \in M$ assign $init(\alpha) \leftarrow \alpha$ and $constr(\alpha) \leftarrow O_\emptyset(C')$
3. **while** there exist $\alpha, \beta \in M$ and different paths $P_1 = \alpha, \gamma_1, \ldots, \beta$, $P_2 = \alpha, \gamma_2, \ldots, \beta$ such that: $O(P_1)$ and $O(P_2)$ are incompatible **and** $init(\delta_1) \neq init(\delta_2)$ for all pairs of internal vertices $\delta_1 \in P_1$ and $\delta_2 \in P_2$ **and** there is no alternative pair $P_1'$ and $P_2'$ of paths with $O(P_1')$ or $O(P_2')$ being a proper subgraph of either $O(P_1)$ or $O(P_2)$
4.     **do** the following:
5.         let $\tau_1 = \alpha \to \gamma_1 \in T$ and $\tau_2 = \alpha \to \gamma_2 \in T$ be transitions corresponding to $P_1$ and $P_2$
6.         $\hat{M} \leftarrow M$, $\hat{T} \leftarrow T$, $\hat{CA} \leftarrow CA$
7.         remove $\alpha$ from $\hat{M}$ and add to $\hat{M}$ two new modes $\alpha_1$ and $\alpha_2$
8.         assign $init(\alpha_1) \leftarrow init(\alpha)$, $init(\alpha_2) \leftarrow init(\alpha)$, $constr(\alpha_1) \leftarrow CA(\tau_1)$, $constr(\alpha_2) \leftarrow CA(\tau_2)$
9.         remove transitions and constraint assignments involving $\alpha$ from $\hat{T}$ and $\hat{CA}$
10.        for each $\tau = \alpha \to \delta \in T$, with $\tau \neq \tau_1, \tau \neq \tau_2$ add transition $\hat{\tau} = \alpha_1 \to \delta$ to $\hat{T}$ and assign $\hat{CA}(\hat{\tau}) \leftarrow CA(\tau)$
11.        for each $\tau = \alpha \to \delta \in T$, with $\tau \neq \tau_1, \tau \neq \tau_2$ add transition $\hat{\tau} = \alpha_2 \to \delta$ to $\hat{T}$ and assign $\hat{CA}(\hat{\tau}) \leftarrow CA(\tau)$
12.        add transition $\hat{\tau} = \alpha_1 \to \gamma_1$ to $\hat{T}$ and assign $\hat{CA}(\hat{\tau}) \leftarrow CA(\tau_1)$
13.        add transition $\hat{\tau} = \alpha_2 \to \gamma_2$ to $\hat{T}$ and assign $\hat{CA}(\hat{\tau}) \leftarrow CA(\tau_2)$
14.        for each $\tau = \delta \to \alpha \in T$ if $CA(\tau)$ and $CA(\tau_1)$ are consistent then add transition $\hat{\tau} = \delta \to \alpha_1$ to $\hat{T}$ and assign $\hat{CA}(\hat{\tau}) \leftarrow CA(\tau) \cup CA(\tau_1)$
15.        for each $\tau = \delta \to \alpha \in T$ if $CA(\tau)$ and $CA(\tau_2)$ are consistent then add transition $\hat{\tau} = \delta \to \alpha_2$ to $\hat{T}$ and assign $\hat{CA}(\hat{\tau}) \leftarrow CA(\tau) \cup CA(\tau_2)$
16.        $M \leftarrow \hat{M}$, $T \leftarrow \hat{T}$, $CA \leftarrow \hat{CA}$
17. **return** $G(\mathcal{F}, CA)$

Algorithm *RefineStateSpace* is greedy and heuristic by its nature and its output depends from the order in which pairs of paths $P_1$ and $P_2$ are selected in *Step 3*. Nevertheless we can guarantee that the algorithm terminates and outputs state space graph of constrained frame $(\mathcal{F}, CA)$ .

Let us denote by $n(\mathcal{H})$ and $m(\mathcal{H})$ correspondingly the number of modes and the number of transitions in $\mathcal{H}$. Let us denote by $N(\mathcal{H})$ the number of different complete (for each of the variables) orderings of transition constants of $\mathcal{H}$ that are consistent with $Min(\mathcal{F}(\mathcal{H})) \cup External(\mathcal{H})$.

**Proposition 1.** *Algorithm RefineStateSpace terminates after a finite number of steps and outputs a state space graph $G$ with at most $n(\mathcal{H}) \times N(\mathcal{H})$ vertices.*

*Proof.* For each vertex of initial graph $\alpha \in V(G_0)$ consider set $S_\alpha = \{\beta \in V(G) | init(\beta) = \alpha\}$. By the design of the algorithm for each pair of different vertices $\beta_1, \beta_2 \in S_\alpha$ constraints $constr(\beta_1)$ and $constr(\beta_2)$ are incompatible. Thus there are at most $N(\mathcal{H})$ vertices in each such set $S_\alpha$ and therefore at most $n(\mathcal{H}) \times N(\mathcal{H})$ vertices in $V(G)$.

The straightforward implementation of each **while** step of the algorithm has $O(N(\mathcal{H})(n(\mathcal{H}) + m(\mathcal{H})))$ time, leading to comparatively high total time complexity $O((N(\mathcal{H})(n(\mathcal{H}) + m(\mathcal{H})))^2)$. However it is not difficult, although technically somewhat involved, to provide implementation with $O(N(\mathcal{H})n(\mathcal{H})(n(\mathcal{H}) + m(\mathcal{H})))$ running time, which compares well with running of state space analysis separately for each of $N(\mathcal{H})$ complete orderings and *analyzing* the whole set of $N(\mathcal{H})$ graphs afterwards.

The way in which algorithm constructs $(\mathcal{F}, CA)$ from initial $(\mathcal{F}_0, CA_0)$ closely resembles the way of defining a constrained frame refinement – essentially the algorithm explicitly constructs the mode and transition mappings $m$ and $t$. Moreover, when mode $\alpha$ is split into modes $\alpha_1$ and $\alpha_2$ each path involving transition $\alpha \to \delta$ is preserved by replacing $\alpha$ either with $\alpha_1$ or $\alpha_2$. Thus, if all runs of frame $(\mathcal{F}_0, CA_0)$ are infinite we have $(\mathcal{F}, CA) < (\mathcal{F}_0, CA_0)$. For finite runs however there is a possibility in each step to lose the first mode of the sequence. In most cases this will not be a problem, since all SCCs, apart from the ones consisting of a single mode, will be preserved, however the algorithm can be adjusted to preserve such modes.

**Proposition 2.** *If there are no finite runs of initial frame $(\mathcal{F}_0, CA_0)$ then for $G(\mathcal{F}, CA)$ computed by algorithm RefineStateSpace the following holds:*

1. *For each complete (for each of the variables) ordering $Ord$ of transition constants and an attractor $S$ in state space graph $G(\mathcal{F}_0, CA_{Ord})$ there is an attractor $\hat{S}$ in $G(\mathcal{F}, CA)$ with the same number of vertices and preserving all paths in $S$.*
2. *For each attractor $S$ in $G(\mathcal{F}, CA)$ there is a complete ordering $Ord$ with state space graph $G(\mathcal{F}_0, CA_{Ord})$ containing attractor $\hat{S}$ with the same number of vertices and preserving all paths in $S$.*

*Proof.* 1. We have already shown that all infinite paths and thus the whole attractor $\hat{S}$ will be preserved in $G(\mathcal{F}, CA)$. Let $\hat{S}$ be one of the attractors with minimal number of modes such that subset $A \subseteq \hat{S}$ is mapped to $S$. If there exists a non-mapped mode $\alpha \in \hat{S}$ and $\alpha \notin A$ then due to minimality of $\hat{S}$ constraints for all transitions to $\alpha$ must be compatible with $Ord$, thus we should have $\alpha \in A$.

2. Consider an infinite path $P$ in $\hat{S}$ and constraint $O(P)$. Then for every complete ordering $Ord$ compatible with the $O(P)$ there will be a corresponding attractor $S$ in $G(\mathcal{F}_0, CA_{Ord})$ to which $\hat{S}$ can be mapped. Clearly all modes of $S$ should also have preimages in $\hat{S}$.

# 5   Computational Experiments

For the toy example in Fig. 1 we have number of modes $n(\mathcal{H}) = 8$ and number of different orderings $N(\mathcal{H}) = 6$. By performing analysis of space graphs for all the 6 orderings we find 5 different attractors: $A_1$ (8 states), $A_2$ (4 states), $A_3$, $A_4$ and $A_5$ (each with 2 states). The state graphs for 6 possible orderings correspondingly contain the following sets of attractors: $\{A_1\}$, $\{A_2\}$, $\{A_2, A_3\}$ (for two different orderings), $\{A_4\}$ and $\{A_4, A_5\}$.

When we use algorithm *RefineStateSpace* for analysis of this model, we find the same sets of attractors. In principle number of states in refined graph depends on the order in which the pairs of paths is chosen by the algorithm. However for this example the algorithm consistently produced graphs with 40 states (this can be shown to be the smallest possible) in which only attractor $A_3$ was present in 2 copies (a separate one for each of the orderings allowing $A_3$).

The constrained HSM frame of our $\lambda$-phage model is derived from the earlier model given in FSLM formalism. The model is described in [5] and [13] and has been derived from semi-formal yet very well developed biological model of $\lambda$-phage from [11].

The model includes 11 genes: *N, cI, cII, cIII, cro, xis, int, O, P, Q* and an artificial gene *Struc* that stands for all genes producing structural proteins. The activity of these genes is regulated by 10 binding sites, 4 of them each can bind one and other 6 can bind two different proteins. The initial constrained HSM frame $(\mathcal{F}(\mathcal{H}) = \langle M, X, C', T, MF' \rangle, Min(\mathcal{F}(\mathcal{H})) \cup External(\mathcal{H}))$ thus contains $3^6 \times 2^4 = 11664$ modes in $M$ corresponding to all the possible combinations of the binding site states. Each of these modes has between 10 and 16 outgoing transitions in $T$ – there is at least one transition for each binding site representing the change of its state and two outgoing transitions for each site binding two proteins, if this site is unoccupied in this particular mode. The set $X$ contains 11 variables corresponding to the number of genes and there are 32 constants in $C'$ for binding affinities (see [5,13]). Mode-function assignments $MF'$ are derived from FSLM model by replacing concrete linear growth/degradation functions by values from $\{\nearrow, \searrow, \rightarrow\}$.

Constraint assignments $Min(\mathcal{F}(\mathcal{H}))$ are defined by $\mathcal{F}(\mathcal{H})$ and there are 42 orderings of binding affinities that are consistent with known biological facts that define $External(\mathcal{H})$ (in the most compact form these biological constraints are described in [6]). Finally, from the frame are removed transitions for which constraints assigned by $External(\mathcal{H})$ and $Min(\mathcal{F}(\mathcal{H}))$ are incompatible.

Additionally we can remove all modes that are not reachable under any of 42 allowed threshold orderings (e.g. if for some gene $G$ there are two binding sites: $b_1$ with dissociation constant $c_1$ and $b_2$ with association constant $c_2$, and according to $External(\mathcal{H})$ we should have $c_2 < c_1$, then state in which $b_1$ is occupied but $b_2$ is free is not reachable). This reduces number of modes in $\mathcal{F}(\mathcal{H})$ to 2890 (the removal of these modes does not change the attractors found by algorithm *RefineStateSpace*).

Thus for our $\lambda$-phage model we have $n(\mathcal{H}) = 2890$ and $N(\mathcal{H}) = 42$. In [6] we have already shown that for each $Ord$ of 42 orderings consistent with $External(\mathcal{H})$ all the state spaces of $G(\mathcal{F}(\mathcal{H}), CA_{Ord})$ contain only 2 attractors that correspond to *lysis* (12 states) and *lysogeny* (2 states) behaviors of $\lambda$-phage (for details about the structure of these attractors see [6]).

When instead of analyzing each consistent ordering $Ord$ separately we applied algorithm *RefineStateSpace* to constrained frame $(\mathcal{F}(\mathcal{H}), Min(\mathcal{F}(\mathcal{H})) \cup External$ $(\mathcal{H})$ it produced graph with 19693 states. In this case $n(\mathcal{H}) \times N(\mathcal{H}) = 121380$, so we needed to analyze around 6 times fewer states compared to individual analysis of space state graphs for all 42 orderings. Also, the total number of attractors was proportionally smaller – on average each of the attractors found was shared by 7 different orderings.

# 6    Conclusions and Discussion

Whilst it is known that in general cases the method we propose here may not be able to produce non-trivial results (the whole state space can consist of just a single attractor), we have shown that it can be useful for analysis of specific models, including the relatively complex model of $\lambda$-phage. The results for the few existing models are encouraging – all the attractors we have found by analysis of qualitative information incorporated in frames describe behaviors that can be achieved by underlying HSM with explicitly defined quantitative parameter values. Moreover an interesting observation is that the same attractor structure can be shared by many different orderings of transition constants, even when the complete state space graphs for these orderings are different.

Regarding future work, firstly there is a range of questions concerning the formalism that we have developed here and algorithms for state space analysis. E.g. what can we say about general mathematical structure of frame refinements? Can we design a good heuristic for *RefineStateSpace* algorithm that will minimize the number of states in constructed frame refinement (or even, can we design an efficient algorithm that always computes a refinement with the minimal number of states)? Here we have not concentrated much on these questions, partially because the underlying mathematical formalism might still be adjusted and it is not clear as yet which parts of it will be really essential for analysis of HSM representing real biological systems.

Secondly, there are questions how our approach can be extended to better answer questions of biological nature. For instance, we have already indicated the objective to include in models additional information about comparative growth rates of growth/degradation functions (there are examples that show that such inclusion may be useful). Another question of biological significance is the identification and study of constraints that are crucial for directing the system's behavior towards a specific attractor – are such constraints consistent over the whole state space, and can we link such constraints to known biological events within models of real biological systems?

# References

1. Ahmad, J., Bernot, J., Comet, J., Lime, D., Roux, O.: Hybrid modelling and dynamical analysis of gene regulatory networks with delays. Complexus **3**, 231–251 (2007)
2. Alur, R., Belta, C., Ivančić, F., Kumar, V., Mintz, M., Pappas, G.J., Rubin, H., Schug, J.: Hybrid modeling and simulation of biomolecular networks. In: Di Benedetto, M.D., Sangiovanni-Vincentelli, A.L. (eds.) HSCC 2001. LNCS, vol. 2034, pp. 19–32. Springer, Heidelberg (2001)
3. Bartocci, E., Liò, P., Merelli, E., Paoletti, N.: Multiple verification in complex biological systems: the bone remodelling case study. In: Priami, C., Petre, I., de Vink, E. (eds.) Transactions on Computational Systems Biology XIV. LNCS, vol. 7625, pp. 53–76. Springer, Heidelberg (2012)
4. Batt, G., Ben Salah, R., Maler, O.: On timed models of gene networks. In: Raskin, J.-F., Thiagarajan, P.S. (eds.) FORMATS 2007. LNCS, vol. 4763, pp. 38–52. Springer, Heidelberg (2007)
5. Brazma, A., Schlitt, T.: Reverse engineering of gene regulatory networks: a finite state linear model. Genome Biol. **4**(P5), 1–31 (2003)
6. Brazma, R., Cerans, K., Ruklisa, D., Schlitt, T., Viksna, J.: HSM - a hybrid system based approach for modelling intracellular networks. Gene **518**, 70–77 (2013)
7. de Jong, H., Gouze, J., Hernandez, C., Page, M., Sari, T., Geiselmann, J.: Qualitative simulation of genetic regulatory networks using piecewise-linear models. Bull. Math. Biol. **66**, 301–340 (2004)
8. Fromentin, J., Eveillard, D., Roux, O.: Hybrid modeling of biological networks: mixing temporal and qualitative biological properties. BMC Syst. Biol. **4**(79), 11 (2010)
9. Ghosh, R., Tomlin, C.: Symbolic reachable set computation of piecewise affine hybrid automata and its application to biological modelling: Delta-notch protein signalling. Syst. Biol. **1**, 170–183 (2004)
10. Grosu, R., Batt, G., Fenton, F.H., Glimm, J., Le Guernic, C., Smolka, S.A., Bartocci, E.: From cardiac cells to genetic regulatory networks. In: Gopalakrishnan, G., Qadeer, S. (eds.) CAV 2011. LNCS, vol. 6806, pp. 396–411. Springer, Heidelberg (2011)
11. McAdams, H., Shapiro, L.: Circuit simulation of genetic networks. Science **269**, 650–656 (1995)
12. Ruklisa, D., Brazma, A., Viksna, J.: Reconstruction of gene regulatory networks under the finite state linear model. Genome Inform. **16**, 225–236 (2005)
13. Schlitt, T., Brazma, A.: Modelling in molecular biology: describing transcription regulatory networks at different scales. Philos. Trans. R. Soc. Lond. B **361**, 483–494 (2006)
14. Serra, R., Vilani, M., Barbieri, A., Kaufmfman, S., Colacci, A.: One the dynamics of random boolean networks subject to noise: attractors, ergodic sets and cell types. J. Theor. Biol. **265**, 185–193 (2010)
15. Siebert, H., Bockmayr, A.: Temporal constraints in the logical analysis of regulatory networks. Theor. Comput. Sci. **391**, 258–275 (2008)
16. Thomas, D., Thieffry, R., Kaufman, M.: Dynamic behaviour of biological regulatory networks. i. biological role of feedback loops and practical use of the concept of the loop-characteristic state. Bull. Math. Biol. **57**, 247–276 (1995)
17. Thomas, D., Thieffry, R., Kaufman, M.: Dynamic behaviour of biological regulatory networks. ii. immunity control in bacteriophage lamda. Bull. Math. Biol. **57**, 277–297 (1995)

# Parameter Synthesis Using Parallelotopic Enclosure and Applications to Epidemic Models

Thao Dang[1], Tommaso Dreossi[1,2(✉)], and Carla Piazza[2]

[1] VERIMAG, 2 Avenue de Vignate, 38610 Gieres, France
{thao.dang,tommaso.dreossi}@imag.fr
[2] Univerisity of Udine, via delle Scienze 206, 33100 Udine, Italy
carla.piazza@uniud.it

**Abstract.** We consider the problem of refining a parameter set to ensure that the behaviors of a dynamical system satisfy a given property. The dynamics are defined through parametric polynomial difference equations and their Bernstein representations are exploited to enclose reachable sets into parallelotopes. This allows us to achieve more accurate reachable set approximations with respect to previous works based on axis-aligned boxes. Moreover, we introduce a symbolical precomputation that leads to a significant improvement on time performances. Finally, we apply our framework to some epidemic models verifying the strength of the proposed method.

**Keywords:** Parameter synthesis · Polynomial systems · Bernstein basis · Symbolic computation · Epidemic models

## 1 Introduction

This work deals with the following problem: given a dynamical system with uncertain parameters, find a parameter set which guarantees that all the possible simulations of the model satisfy a desired property. The dynamics of the considered system are defined as discrete-time polynomials, the sets reachable by the system are represented with *parallelotopes* (the $n$-dimensional generalization of parallelograms), the parameter sets are represented by polytopes, and the desired property is specified in terms of a linear inequality.

The technique proposed in this paper advances our previous results on parameter synthesis [1] and reachability analysis of polynomial systems [2–4], both based on representation of polynomials in the Bernstein form [5]. Here we introduce a more precise representation of the state of the system and we develop a faster algorithm to synthesize the parameters and compute the reachable sets. The first contribution consists in changing the representation of reachable sets from axis-aligned boxes to parallelotopes. The second relies on the introduction of a precomputation on the dynamics of the system that allows us to save

This work is partially supported by Istituto Nazionale di Alta Matematica (INdAM) and the project MALTHY (ANR-13-INSE-003).

© Springer International Publishing Switzerland 2015
O. Maler et al. (Eds.): HSB 2013 and 2014, LNBI 7699, pp. 67–82, 2015.
DOI: 10.1007/978-3-319-27656-4_4

calculations during the synthesis of the parameters and the simulation of the model.

We apply our technique on some epidemic models, a class of biological systems representing the evolution of infectious maladies. These models are helpful to make predictions on disease spread and their study can aid the planning of strategies aimed to reduce the effects of possible future outbreaks. In this context, the parameter synthesis is of particular interest, since it allows us to reason on specific plans and their effect on the population. For instance, as we will show in Sect. 5, particular isolation and treatment policies may have significant effects on the infection and mortality rates.

Parameter synthesis has been considered using various optimization based techniques. Model checking [6–9] and guided simulation [10] have been proposed to analyze the parameters of biochemical models and identify parameter values that falsify a property. The closest work to ours are [11,12] in which the parameter sets are represented as boxes and the reachable sets are approximated via sensitivity analysis. The main difference here relies in a more compact representation based on polytopes and in the use of linear constraints in the refinement process.

The paper is organized as follows. In Sect. 2 we introduce some definitions and we state the problem. Sections 3 and 4 are dedicated to the description of the algorithms to compute the evolution of the system and to synthesize its parameters. In Sect. 5 we apply our technique on three epidemic models (SIR, SARS and Influenza), showing the improvements of the new technique with respect to the previous. Finally, we conclude in Sect. 6 with a brief discussion.

## 2    Preliminaries

We consider a parametric discrete-time dynamical system described by

$$\begin{aligned} \mathbf{x}(k+1) &= f(\mathbf{x}(k), \mathbf{p}) \\ \mathbf{x}(0) &\in X^0 \end{aligned} \tag{1}$$

where $\mathbf{x} \in \mathbb{R}^n$ is the vector of state variables ($\mathbb{R}$ denotes the set of reals), $\mathbf{p} \in P \subseteq \mathbb{R}^m$ is the vector of uncertain parameters, $f$ is a vector of $n$ multi-variate polynomials of the form $f_i : \mathbb{R}^n \times \mathbb{R}^m \to \mathbb{R}$ for each $i \in \{1, \ldots, n\}$. The set $X^0 \subseteq \mathbb{R}^n$ is called the *initial set*. The set $P$ is called the *initial parameter set*.

Given an initial set $X^0$, at each step the set of all the states visited by the dynamical system (1) can be computed as the solution of the recursion $X^{j+1} = \{f(\mathbf{x}, \mathbf{p}) \mid \mathbf{x} \in X^j, \mathbf{p} \in P\}$, for $j = 0, 1, \ldots, K$. With the notation $\mathcal{R}_P^K$ we emphasize that the reachable set is computed for the fixed parameter set $P$. This calculation at each step amounts to computing the parametric image of a set through the polynomial $f$. This is the core problem we address before proceeding to the parameter synthesis algorithm. Let us formally state such image computation problem.

*Problem 1 (Image computation).* Let $f : \mathbb{R}^n \times \mathbb{R}^m \to \mathbb{R}^n$, $X \subseteq \mathbb{R}^n$ and $P \subseteq \mathbb{R}^m$. We are interested in computing the image of $X \times P$ through $f$, that is the set

$$f(X, P) = \{(f_1(\mathbf{x}, \mathbf{p}), \dots, f_n(\mathbf{x}, \mathbf{p})) \mid \mathbf{x} \in X, \mathbf{p} \in P\}.$$

We will base the iterative computation of $X^j$ on the Bernstein representation of polynomials [13], which we recall in the following section. Once the image computation problem is solved, we will focus on the problem of constraining the parameter set $P$ so that the resulting system satisfies a safety property.

*Problem 2 (Parameters refinement).* Let $\mathcal{F} = \{\mathbf{x} \mid s(\mathbf{x}) \geq 0\} \subseteq \mathbb{R}^n$ be an unsafe set where $s : \mathbb{R}^n \to \mathbb{R}$ is a linear constraint over the state variables. We are interested in finding the largest subset $P_s \subseteq P$ such that starting from the initial set $X^0$, the system does not enter the unsafe set $\mathcal{F}$ up to time $K$, that is

$$\forall \mathbf{p} \in P_s \ \forall j \in \{0, 1, \dots, K\} \ \forall \mathbf{x} \in \mathcal{R}^j_{P_s} : s(\mathbf{x}) < 0.$$

## The Bernstein Basis for Polynomials

A multi-index is a vector $\mathbf{i} = (i_1, i_2, \dots, i_n)$ where each $i_j$ is a non-negative integer. Given two multi-indexes $\mathbf{i}$ and $\mathbf{d}$, we write $\mathbf{i} \leq \mathbf{d}$ ($\mathbf{d}$ dominates $\mathbf{i}$) if for all $j \in \{1, \dots, n\}$, $i_j \leq d_j$. Also, we write $\mathbf{i}/\mathbf{d}$ for $(i_1/d_1, i_2/d_2, \dots, i_n/d_n)$ and $\binom{\mathbf{d}}{\mathbf{i}}$ for the product of binomial coefficients $\binom{d_1}{i_1}\binom{d_2}{i_2}\dots\binom{d_n}{i_n}$. Moreover, we use $\mathcal{B}^n$ to denote the $n$-dimensional unit box $[0, 1]^n \subseteq \mathbb{R}^n$.

A parametric polynomial $\rho : \mathbb{R}^n \times \mathbb{R}^m \to \mathbb{R}$ can be represented using the power basis as follows:

$$\rho(\mathbf{x}, \mathbf{p}) = \sum_{\mathbf{i} \in I^\rho} \mathbf{a_i}(\mathbf{p}) \mathbf{x^i}$$

where $\mathbf{i} = (i_1, i_2, \dots, i_n)$ is a multi-index of size $n$ and $\mathbf{x^i}$ denotes the monomial $x_1^{i_1} x_2^{i_2} \cdots x_n^{i_n}$. The set $I^\rho$ is called the multi-index set of $\rho$. The *degree* $\mathbf{d}$ of $\rho$ is the smallest multi-index which dominates all the multi-indexes of $I^\rho$ (i.e., $\forall \mathbf{i} \in I^\rho : \mathbf{i} \leq \mathbf{d}$). The coefficients $\mathbf{a_i}(\mathbf{p})$ are functions of the parameters $\mathbf{p}$ of the form $\mathbb{R}^m \to \mathbb{R}$.

Bernstein basis polynomials of degree $\mathbf{d}$ is a basis for the space of polynomials of degree at most $\mathbf{d}$ over $\mathbb{R}^n$. In particular, for $\mathbf{x} = (x_1, \dots, x_n) \in \mathbb{R}^n$, the $\mathbf{i}^{th}$ Bernstein polynomial of degree $\mathbf{d}$ is defined as $\mathcal{B}_{\mathbf{d},\mathbf{i}}(\mathbf{x}) = \beta_{d_1,i_1}(x_1) \dots \beta_{d_n,i_n}(x_n)$ where for a real number $y$, $\beta_{d_j,i_j}(y) = \binom{d_j}{i_j} y^{i_j} (1 - y)^{d_j - i_j}$. Hence, the polynomial $\rho$ can also be represented using Bernstein basis and it can be written as

$$\rho(\mathbf{x}, \mathbf{p}) = \sum_{\mathbf{i} \in I^\rho} \mathbf{b_i}(\mathbf{p}) \mathcal{B}_{\mathbf{d},\mathbf{i}}(\mathbf{x})$$

where for each $\mathbf{i} \in I^\rho$ the Bernstein coefficient $\mathbf{b_i}(\mathbf{p})$ is

$$\mathbf{b_i}(\mathbf{p}) = \sum_{\mathbf{j} \leq \mathbf{i}} \frac{\binom{\mathbf{i}}{\mathbf{j}}}{\binom{\mathbf{d}}{\mathbf{j}}} \mathbf{a_j}(\mathbf{p}).$$

Bernstein representation is of particular interest due to useful geometric properties of its coefficients. If we refer to the unit box $\mathcal{B}^n$, Bernstein representation can be used to bound $\rho$, since:

$$\forall \mathbf{x} \in \mathcal{B}^n \ \forall \mathbf{p} \in P : \ \rho(\mathbf{x}, \mathbf{p}) \in [m^\rho, M^\rho] \tag{2}$$

where $m^\rho = min\{\mathbf{b_i}(\mathbf{p}) \mid \mathbf{i} \in I^\rho \wedge \mathbf{p} \in P\}$ and $M^\rho = max\{\mathbf{b_i}(\mathbf{p}) \mid \mathbf{i} \in I^\rho \wedge \mathbf{p} \in P\}$. We can see here the advantage of Bernstein representation in the analysis of parametric systems as it succinctly captures the bounds of the reachable set.

## 3   Image Over-Approximation

### 3.1   Parallelotope Representations

In [1] a method to step-wise over-approximate the image computation through axis-aligned boxes has been proposed. Here we extend the method to parallelotopes, i.e., the $n$-dimensional generalizations of parallelepipeds. The use of parallelotopes makes the method more flexible as far as the choice of the initial set $X^0$ is concerned and it allows one to obtain better approximations.

A parallelotope $X$ is a centrally symmetric convex polyhedron that can be described using the *generator representation* as follows.

**Definition 1 ($\mathcal{P}_{gen}(\mathbf{q}, G)$).** *Let $G = \{\mathbf{g_1}, \ldots, \mathbf{g_n}\}$ be a set of $n$ linearly independent vectors in $\mathbb{R}^n$ and $\mathbf{q}$ be a point in $\mathbb{R}^n$. The parallelotope $X$ generated by $G$ and $\mathbf{q}$ is:*

$$X = \mathcal{P}_{gen}(\mathbf{q}, G) = \{\mathbf{q} + \sum_{j=1}^{n} \alpha_j \mathbf{g}_j \mid (\alpha_1, \ldots, \alpha_n) \in \mathcal{B}^n \wedge \mathbf{g}_j \in G\}.$$

The vectors $\mathbf{g}_j$ are called *generators* of the parallelotope and $\mathbf{q}$ is called *base vertex*. Given a set of generators $G = \{\mathbf{g_1}, \ldots, \mathbf{g_n}\}$ and a base vertex $\mathbf{q}$, we will also represent the parallelotope generated by $G$ and $\mathbf{q}$ through the notation

$$X = \mathcal{P}_{gen}(\mathbf{q}, G) = \{\gamma_{(\mathbf{q}, G)}(\alpha) \mid \alpha \in \mathcal{B}^n\}$$

where $\alpha = (\alpha_1, \ldots, \alpha_n)$ and $\gamma_{(\mathbf{q}, G)}$ is the linear function defined as

$$\gamma_{(\mathbf{q}, G)}(\alpha) = \mathbf{q} + \sum_{j=1}^{n} \alpha_j \mathbf{g}_j.$$

Such notation emphasizes the aspect that a parallelotope can be thought as the affine transformation of the unit box $\mathcal{B}^n$.

**Definition 2 ($\mathcal{P}_{con}(\Lambda, \mathbf{d})$).** *Let $\Lambda$ be a $2n \times n$ matrix such that $\Lambda = (\Lambda_j)_{j=1,\ldots,2n}$ and $\forall j \in \{1, \ldots, n\}\Lambda_j = -\Lambda_{j+n}$ and let $\mathbf{d} \in \mathbb{R}^{2n}$. The parallelotope $X$ generated by $\Lambda$ and $\mathbf{d}$ is:*

$$X = \mathcal{P}_{con}(\Lambda, \mathbf{d}) = \{\mathbf{x} \mid \Lambda \mathbf{x} \leq \mathbf{d}\}.$$

The above representation is called the *constraint representation*. The rows of the matrix $\Lambda$ are called *directions* and the vector $\mathbf{d} = (d_1, \ldots, d_{2n}) \in \mathbb{R}^{2n}$ is called the *offset*. If the direction matrix $\Lambda$ is fixed, $\Lambda$ is called the *template matrix*. In this case the paralellotopes are a special case of template polyhedra [14].

Note that the Bernstein representation presented in the previous section allows us to easily bound the values of a polynomial over the unit-box (see Eq. 2). Hence, the generator representation of $X$, which can be interpreted as a function of $\alpha \in \mathcal{B}^n$, is suitable to over-approximate $f(X, P)$. On the other hand, we will exploit the constraint representation to compute a new parallelotope which over-approximates such an image.

Let us now focus on the image computation problem. Let $X$ be a parallelotope represented through its generator representation $X = \mathcal{P}_{gen}(\mathbf{q}, G)$ and $\Lambda$ be a template matrix. We are interested in computing a parallelotope $X' = \mathcal{P}_{con}(\Lambda, \mathbf{d})$ such that $f(X, P) \subseteq X'$. More concretely, we want to determine the offset $\mathbf{d} \in \mathbb{R}^{2n}$ such that $f(\mathcal{P}_{gen}(\mathbf{q}, G), P) \subseteq \mathcal{P}_{con}(\Lambda, \mathbf{d})$. The following Lemma shows how to determine such an offset $\mathbf{d}$.

**Lemma 1.** *Let $\mathbf{d} = (d_1, \ldots, d_{2n})$ be such that, for each $j \in \{1, \ldots, 2n\}$, the inequality $d_j \geq max\{\Lambda_j f(\mathbf{y}, \mathbf{p}) \mid \mathbf{y} \in \mathcal{P}_{gen}(\mathbf{q}, G) \wedge \mathbf{p} \in P\}$ holds. The inclusion $f(\mathcal{P}_{gen}(\mathbf{q}, G), P) \subseteq \mathcal{P}_{con}(\Lambda, \mathbf{d})$ is guaranteed.*

Exploiting the generator representation, the above condition can be rewritten as $d_j \geq max\{h^j(\alpha, \mathbf{p}) \mid \alpha \in \mathcal{B}^n \wedge \mathbf{p} \in P\}$, where $h_j(\alpha, \mathbf{p}) = \Lambda_j f(\gamma_{(\mathbf{q}, G)}(\alpha), \mathbf{p})$. It is not hard to see that $h_j(\alpha, \mathbf{p})$ is a polynomial function of $\alpha$ and its coefficients are linear functions of the parameters $\mathbf{p}$. Furthermore, the domain we are interested in is exactly the unit box; therefore we can straightforwardly apply the Bernstein representation to compute an upper bound of the function $h_j(\alpha, \mathbf{p})$ with $\alpha \in \mathcal{B}^n$.

We denote the set of the Bernstein coefficients of $h_j(\alpha, \mathbf{p})$ as $B^{h_j}(\mathbf{p}) = \{\mathbf{b}_{\mathbf{i}}^{h_j}(\mathbf{p}) \mid \mathbf{i} \in I^{h_j}\}$. Here we write each Bernstein coefficient as a function of $\mathbf{p}$ because they are computed from monomial coefficients which are linearly dependent on the parameters $\mathbf{p}$.

**Theorem 1.** *Let $\mathbf{d} = (d_1, \ldots, d_{2n})$ be such that for each $j \in \{1, \ldots, 2n\}$ the component $d_j$ is defined as $d_j = max\{\mathbf{b}_{\mathbf{i}}^{h_j}(\mathbf{p}) \mid \mathbf{i} \in I^{h_j} \wedge \mathbf{p} \in P\}$. The vector $\mathbf{d}$ satisfies the inclusion $f(\mathcal{P}_{gen}(\mathbf{q}, G), P) \subseteq \mathcal{P}_{con}(\Lambda, \mathbf{d})$.*

## 3.2  Bounding Reachable Sets

In order to formalize the algorithm for bounding the reachable set, we rewrite the generator representation of parallelotopes explicitly distinguishing between the directions of the generators and their lengths.

Let $G = \{\mathbf{g}_1, \mathbf{g}_2, \ldots, \mathbf{g}_n\}$ be a set of generators. Let $\beta_i \in \mathbb{R}$ be the Euclidian norm of $\mathbf{g}_i$ and $\mathbf{u}_i$ be the versor (vector of norm 1) of $\mathbf{g}_i$, i.e., $\mathbf{g}_i = \beta_i \mathbf{u}_i$. Let $\beta = (\beta_1, \beta_2, \ldots, \beta_n)$ and $U = \{\mathbf{u}_1, \mathbf{u}_2, \ldots, \mathbf{u}_n\}$. With a slight abuse of notation, the generator representation can be rewritten as $\mathcal{P}_{gen}(\mathbf{q}, \beta, U) = \{\gamma_U(\alpha, \mathbf{q}, \beta) \mid \alpha \in \mathcal{B}^n\}$, where $\gamma_U(\alpha, \mathbf{q}, \beta)$ is the linear function in $\alpha$ defined as

$$\gamma_U(\alpha, \mathbf{q}, \beta) = \mathbf{q} + \sum_{j=1}^{n} \alpha_j \beta_j \mathbf{u}_j.$$

When we work on parallelotopes using the constraint representation we can fix a template matrix $\Lambda$ and let the offset $\mathbf{d}$ be free. In this way we symbolically denote an infinite set of parallelotopes. On the generator representation this corresponds to the choice of a set $U$ of $n$ versors, while the norms $\beta$ are free.

We focus on a single reachability step: given a parallelotope $X = \mathcal{P}_{gen}(\mathbf{q}, \beta, U)$, we want to compute $\mathbf{q}'$ and $\beta'$ such that $\mathcal{P}_{gen}(\mathbf{q}', \beta', U)$ over-approximates the set $f(X, P)$. The set $f(X, P)$ can be characterized as

$$f(X, P) = f(\gamma_U(\mathcal{B}^n, \mathbf{q}, \beta), P) = \{f(\gamma_U(\alpha, \mathbf{q}, \beta), \mathbf{p}) \mid \alpha \in \mathcal{B}^n \wedge \mathbf{p} \in P\}.$$

Hence, if we find $\mathbf{q}'$ and $\beta'$ such that $f(\gamma_U(\mathcal{B}^n, \mathbf{q}, \beta), P) \subseteq \gamma_U(\mathcal{B}^n, \mathbf{q}', \beta')$, we obtain an over-approximation in generator representation of $f(X, P)$. Such $\mathbf{q}'$ and $\beta'$ can be found passing through an intermediate constraint representation. Let $\Lambda$ be the template matrix which corresponds to the versor generators $U$ (see Sect. 3.3), we try to find the offset $\mathbf{d}$ such that $f(\gamma_U(\mathcal{B}^n, \mathbf{q}, \beta), P) \subseteq \mathcal{P}_{con}(\Lambda, \mathbf{d})$ and then we convert the constraint representation $\mathcal{P}_{con}(\Lambda, \mathbf{d})$ to its generator representation $\mathcal{P}_{gen}(\mathbf{q}', \beta', U)$. The offset $\mathbf{d}$ can be calculated exploiting Theorem 1, i.e., for each $j \in \{1, \ldots, 2n\}$, over-approximating $\max\{\Lambda_j(f(\gamma_U(\alpha, \mathbf{q}, \beta), \mathbf{p})) \mid \alpha \in \mathcal{B}^n \wedge \mathbf{p} \in P\}$, a task that can be carried out taking advantage of the Bernstein representation. Such approach, similarly to the technique described in [1], would require the recomputation of Bernstein coefficients at each reachability step. However, three important aspects have to be pointed out:

1. by definition of the generator representation, once $\mathbf{q}, \beta$, and $\mathbf{p}$ have been chosen, the domain of $\Lambda_j(f(\gamma_U(\alpha, \mathbf{q}, \beta), \mathbf{p}))$ is the unit box $\mathcal{B}^n$, that is exactly the domain on which Bernstein coefficients satisfy Eq. 2;
2. the Bernstein coefficients of the function $\Lambda_j(f(\gamma_U(\alpha, \mathbf{q}, \beta), \mathbf{p}))$ are functions of the form $\mathbf{b_i}(\mathbf{q}, \beta, \mathbf{p})$ linear in $\mathbf{p}$;
3. both the template matrix $\Lambda$ and $U$ are fixed, i.e., at each reachability step the directions of the edges of the parallelotopes are the same.

Since the template matrix $\Lambda$ and $U$ are fixed, we do not need to recompute the Bernestein coefficients of $\Lambda_j(f(\gamma_U(\alpha, \mathbf{q}, \beta), \mathbf{p}))$ at each reachability step but, keeping the parameters $\mathbf{q}, \beta$, and $\mathbf{p}$ symbolically, we can compute them only once obtaining a template of Bernstein coefficients that we evaluate at each reachability step. In the following we formalize this idea.

Given the template matrix $\Lambda \in \mathbb{R}^{2n \times n}$, the set of generator versors $U = \{\mathbf{u}_1, \mathbf{u}_2, \ldots, \mathbf{u}_n\} \subseteq \mathbb{R}^n$ and the dynamics $f : \mathbb{R}^n \times \mathbb{R}^m \rightarrow \mathbb{R}^n$, Algorithm 1 produces a template of Bernstein coefficients $(\Upsilon_j)_{j=1}^{2n}$ (also called control points), that is a $2n$-dimensional vector of vectors of parametrized Bernstein coefficients

---

**Algorithm 1.** Building the Bernstein Coefficients Template

---

1: **function** BUILDTEMPLATE($\Lambda, U, f$)
2:     **for** $j \in \{1, \ldots, 2n\}$ **do**
3:         $h_j \leftarrow \Lambda_j(f(\gamma_U(\alpha, \mathbf{q}, \beta), \mathbf{p}))$
4:         $\Upsilon_j \leftarrow \text{BERNCOEFF}(h_i)$
5:     **end for**
6:     **return** $\Upsilon$
7: **end function**

---

of the form $\mathbf{b}_{j,\mathbf{i}}(\mathbf{q}, \beta, \mathbf{p}) : \mathbb{R}^n \times \mathbb{R}^n \times \mathbb{R}^m \to \mathbb{R}$, where $j \in \{1, \ldots, 2n\}$, and $\mathbf{i} \in I^{f(\gamma_U(\alpha, \mathbf{q}, \beta), \mathbf{p})}$.

Notice that $U$ depends on $\Lambda$, i.e., it can be computed from $\Lambda$. In the next section we will see how this can be done. However, in Algorithm 1 we are not interested in this technical detail and we pass to the function both $\Lambda$ and $U$.

*Example 1.* Let us consider an example of the well known predator-prey Lotka-Volterra model whose two dynamics are $f_1(\mathbf{x}, \mathbf{p}) = x_1 + x_1(a - x_2)$ and $f_2(\mathbf{x}, \mathbf{p}) = x_2 - x_2(c - 2x_1)$. Choosing the generator versors $U = \{(1.0, 0.0), (0.55, 0.83)\}$ and the symbolic base vertex $\mathbf{q} = (q_1, q_2)$, we obtain the generator function and template matrix

$$
\gamma_U(\alpha, \mathbf{q}, \beta) = \begin{pmatrix} q_1 + \beta_1\alpha_1 + \frac{2}{13}(\sqrt{13}\alpha_1\beta_2) \\ q_2 + \frac{3}{13}(\sqrt{13}\alpha_2\beta_2) \end{pmatrix} \qquad \Lambda = \begin{pmatrix} -0.83 & 0.55 \\ 0.00 & -1.00 \\ 0.83 & -0.55 \\ 0.00 & 1.00 \end{pmatrix}
$$

that lead to a collection of functions $h_j$, with $j \in \{1, \ldots, 4\}$, where for instance

$$
h_2 = -(q_2 + \frac{3}{13}(\sqrt{13}\alpha_2\beta_2))(2q_1 - c + 2\alpha_1\beta_1 + \frac{4}{13}(\sqrt{13}\alpha_2\beta_2)).
$$

Finally, from $h_1, h_2, h_3$, and $h_4$ we compute the template coefficients $\Upsilon$ of which, for brevity, we report the first element:

$$
\Upsilon_{1,1} = \mathbf{b}_{1,(0,0)}(\mathbf{q}, \beta, p) = \frac{2\sqrt{13}}{13}q_1(a - q_2)(-\frac{3}{13} - \sqrt{13}q_2(c - q_12)).
$$

At this point, fixing the base vertex $\mathbf{q}$ and the versor norms $\beta$, in order to compute an over-approximation of the reachability step $f(X, P) = f(\gamma_U(\mathcal{B}^n, \mathbf{q}, \beta), P)$, it is sufficient to find the maximum of each row ($j = 1, \ldots, 2n$) of the Bernstein coefficients template over the parameter set $P$. Algorithm 2 formalizes this computation. Each offset $d_j$ of the constraint representation of $f(X, P)$ is derived from the maximum of the $j$-th row of the Bernstein coefficients template $\Upsilon$ over the parameter set $P$ (Line 3). Finally, the constraint representation $\mathcal{P}_{con}(\Lambda, \mathbf{d})$ of the over-approximation of $f(X, P)$ is converted into the generator representation (Line 5). Such conversion (discussed in the next section) computes the essential information to reconstruct the new parallelotope: the new base vertex $\mathbf{q}'$ and the new generator amplitudes $\beta'$.

---
**Algorithm 2.** Bounding the reachable set from $X$
---
1: **function** REACHSTEP($\mathbf{q}, \beta, \Upsilon, P$)
2:     **for** $j \in \{1, \ldots, 2n\}$ **do**
3:         $d_j \leftarrow$ MAX($\mathbf{q}, \beta, \Upsilon_j, P$)
4:     **end for**
5:     **return** $[\mathbf{q}', \beta'] \leftarrow$ CON2GEN($\mathcal{P}_{con}(\Lambda, \mathbf{d})$)
6: **end function**
---

## 3.3   Representation Conversion

We now see how to convert the generator representation of a parallelotope into its constraint representation and vice versa. The switch from generator to constraint representation is useful to compute the best template matrix given a set versors and norms, while the inverse conversion represents the last task in our single step reachability computation (see Algorithm 2, Line 5). The efficiency with which such conversions are performed influences the reachability algorithm performance and thus indirectly the whole parameter synthesis procedure.

**From Generators to Constraints.** Given the generator representation of a parallelotope $\mathcal{P}_{gen}(\mathbf{q}, \beta, U)$ we want to find its equivalent constraint representation $\mathcal{P}_{con}(\Lambda, \mathbf{d})$, i.e., we want to define a function such that given $\mathbf{q}$, $\beta$, and $U$ allows us to compute $\Lambda$ and $d$ such that $\mathcal{P}_{gen}(\mathbf{q}, \beta, U) = \mathcal{P}_{con}(\Lambda, \mathbf{d})$. Let $\mathbf{q} \in \mathbb{R}^n$ be the base vertex, $U = \{\mathbf{u}_1, \ldots, \mathbf{u}_n\}$ be the generator versors set, where for $i = 1, \ldots, n$ it holds $\mathbf{u}_i \in \mathcal{B}^n$, and $\beta = (\beta_1, \beta_2, \ldots, \beta_n) \in \mathbb{R}^n$. Moreover, let $\mathbf{g}_i = \beta_i \mathbf{u}_i$, for $i = 1, \ldots, n$. As first step we calculate the points $\mathbf{p}_1, \ldots, \mathbf{p}_n$ through which we will traverse the hyperplanes of the constraint representation. Such $\mathbf{p}_i$ are obtained adding to the base vertex the unit vectors $\mathbf{u}_i$, that is $\mathbf{p}_i = \mathbf{q} + \mathbf{u}_i$, for $i = 1, \ldots, n$. Let $\pi_i = \mathbf{a}_i \mathbf{x} + d_i$ be the equation of the hyperplane passing through the points $\mathbf{q}, \mathbf{p}_1, \mathbf{p}_2, \ldots, \mathbf{p}_{i-1}, \mathbf{p}_{i+1}, \ldots, \mathbf{p}_n$. The equation $\pi_i$ represents the hyperplane on which lies the $i$-th facet of the parallelotope. The equation $\pi_{i+n} = \mathbf{a}_{i+n} \mathbf{x} + d_{i+n}$ of the hyperplane parallel to $\pi_i$ can be found translating the vertexes used to compute $\pi_i$ by the vector $\mathbf{g}_i$, i.e., $\pi_{i+n}$ is the hyperplane passing through the points $\mathbf{q} + \mathbf{g}_i, \mathbf{p}_1 + \mathbf{g}_i, \mathbf{p}_2 + \mathbf{g}_i, \ldots, \mathbf{p}_{i-1} + \mathbf{g}_i, \mathbf{p}_{i+1} + \mathbf{g}_i, \ldots, \mathbf{p}_n + \mathbf{g}_i$. Let $\underline{d_i}$ and $\overline{d_i}$ be defined as $\underline{d_i} = \min\{d_i, d_{i+n}\}$ and $\overline{d_i} = \max\{d_i, d_{i+n}\}$. Since $\pi_i$ and $\pi_{i+n}$ are parallel, it must hold that $\mathbf{a}_i = \mathbf{a}_{i+n}$. Hence, the portion of the parallelotope included between $\pi_i$ and $\pi_{i+n}$ can be characterized by the inequality $\underline{d_i} \leq \mathbf{a}_i \mathbf{x} \leq \overline{d_i}$ which means that the $i$-th and $(i+n)$-th rows of the template matrix $\Lambda$ are $\Lambda_i = \mathbf{a}_i$ and $\Lambda_{i+n} = -\mathbf{a}_i$, while the $i$-th and $(i+n)$-th directions are $d_i = \overline{d_i}$ and $d_{i+n} = -\underline{d_i}$.

**From Constraints to Generators.** We now see how to compute the opposite conversion. We first rewrite the inequalities given by the template matrix $\Lambda$ and the direction $\mathbf{d}$ in form $-d_{n+i} \leq \Lambda_i \leq d_i$, for $i = 1, \ldots, n$. The base vertex $\mathbf{q}$ and the coordinates of the vertex $\mathbf{v}_i$, for $i = 1, \ldots, n$, that lies on the straight line passing through the $i$-th generator vector applied to the vertex $\mathbf{q}$, are the solution of the linear systems:

$$\begin{pmatrix} \Lambda_1 \\ \vdots \\ \Lambda_n \end{pmatrix} \mathbf{x} = \begin{pmatrix} -d_{n+1} \\ \vdots \\ -d_{2n} \end{pmatrix} \qquad \begin{pmatrix} \Lambda_1 \\ \vdots \\ \Lambda_i \\ \vdots \\ \Lambda_n \end{pmatrix} \mathbf{x} = \begin{pmatrix} -d_{n+1} \\ \vdots \\ d_i \\ \vdots \\ -d_{2n} \end{pmatrix}$$

Hence, the $i$-th generator $\mathbf{g}_i$ is the difference between the vertex $\mathbf{v}_i$ and the base vertex $\mathbf{q}$, i.e., $\mathbf{g}_i = \mathbf{v}_i - \mathbf{q}$. Finally, the versor $\mathbf{u}_i$ and the generator norm $\beta_i$ such that $\mathbf{g}_i = \beta_i \mathbf{u}_i$ are given by $\beta_i = \|\mathbf{g}_i\|$ and $\mathbf{u}_i = \frac{\mathbf{g}_i}{\|\mathbf{g}_i\|}$.

## 4   Parameter Synthesis

In this work we assume that the safety constraint $s$ is linear in $\mathbf{x}$ and that all the coefficients $\mathbf{a_i}$ of the dynamics $f$ are linear in the parameters $\mathbf{p}$. This assumption allows us to reduce the synthesis problem to a set of linear programs.

To check whether the system does not reach the unsafe set $\mathcal{F}$ we can consider the safety function $\sigma = s(f(\gamma_U(\alpha, \mathbf{q}, \beta), \mathbf{p}))$ and its set of Bernstein coefficients $B^\sigma(\mathbf{p}) = \{\mathbf{b_i^\sigma}(\mathbf{p}) \mid \forall \mathbf{i} \in I^\sigma\}$. The following is a sufficient condition for the system $f$ to satisfy the safety property $s$ after one step starting from the set represented by $\gamma_U(\mathcal{B}^n, \mathbf{q}, \beta)$:

$$\forall \mathbf{p} \in P \ \forall \mathbf{i} \in I^\sigma : \ \mathbf{b_i^\sigma}(\mathbf{p}) < 0. \tag{3}$$

Note that since $s$ is a linear function and the parameters $\mathbf{p}$ appear linearly in the dynamics of $f$, the coefficients in the monomial representation of $\sigma$ remain linear in $\mathbf{p}$. This means that the constraints of Eq. 3 are linear inequalities over $\mathbf{p}$. This observation allows us to translate the synthesis problem in the resolution of a linear system of inequalities.

At time $j = 1, \ldots, K$, the parameter set $P^j$ is represented as the solution of the linear system of the form $A^j \mathbf{p} < \mathbf{b}^j$. Before starting the parameter synthesis, we collect in a vector $\Phi$ the Bernstein coefficients of the safety function $\sigma = s(f(\gamma_U(\alpha, \mathbf{q}, \beta), \mathbf{p}))$, keeping symbolically the base vertex $\mathbf{q}$, the generator vector amplitudes $\beta$, and the parameters $\mathbf{p}$. Each element of $\Phi$ is a function $\mathbf{b_i^\sigma}(\mathbf{q}, \beta, \mathbf{p}) : \mathbb{R}^n \times \mathbb{R}^n \times \mathbb{R}^m \to \mathbb{R}$, for all $\mathbf{i} \in I^\sigma$.

Suppose that the state of the system at the $j$-th step is described by the set $X^j$ whose base vertex and generator amplitudes are $\mathbf{q}^j$ and $\beta^j$. The refinement of the parameter set $P^{j-1}$, represented by the system $A^{j-1} \mathbf{p} < \mathbf{b}^{j-1}$, with respect to the constraint $s$ consists of the following steps:

1. for all $\mathbf{b_i^\sigma}(\mathbf{q}, \beta, \mathbf{p}) \in \Phi$ substitute $\mathbf{q}$ and $\beta$ with $\mathbf{q}^j$ and $\beta^j$, respectively. All the functions $\mathbf{b_i^\sigma}(\mathbf{q}^j, \beta^j, \mathbf{p})$ are now linear in $\mathbf{p}$;
2. build the linear system merging $A^{j-1} \mathbf{p} < \mathbf{b}^{j-1}$ with all the constraints $\mathbf{b_i^\sigma}(\mathbf{q}^j, \beta^j, \mathbf{p}) < 0$, for $\mathbf{i} \in I^\sigma$. We denote by $P^j$ the new linear system;

3. check whether $P^j$ has solutions.

If $P^j$ has solutions, i.e., the parameter set is not empty, then the set $X^{j+1} = f(X^j, P^j)$ is safe with respect to the constraint $s$. If $P^j$ has no solution, then either there do not exist parameter values in $P^j$ such that the system can safely evolve, i.e., $\forall \mathbf{p} \in P^j : s(f(X^j, P^j)) \geq 0$, or the over-approximation of the set $X^j$ is inaccurate. In [1] we describe a method to determine whether the parameter set is truly empty and, if it is not the case, how to refine the image computation using a backtracking technique. For simplicity, here we omit this refinement which can be easily introduced also in our new procedure.

The whole parameter synthesis is summarized in Algorithm 3. First, the procedure computes the Bernstein coefficient template $\Upsilon$ and the Bernstein coefficients $\Phi$ of the safety function (Lines 3,4). Then the algorithm enters in a loop that is iterated until either the maximum number of steps $K$ is reached or the parameter set $P^j$ is empty. Each iteration refines the $(j-1)$-th parameter set thanks to the procedure REFPARAMS which exploits the symbolic coefficients stored in $\Phi$ (Line 7). Then, if the refined set $P^j$ is not empty, the algorithm performs a safe single reachability step from the state set $X^j$ with the parameter values $P^j$ (Line 9). As result, the function REACHSTEP returns the base vertex $\mathbf{q}^j$ and the generator amplitudes $\beta^j$ that are the data needed to represent the new reached set $X^j$.

---

**Algorithm 3.** Parameter synthesis w.r.t. $s$

---

1: **function** PARASYNTH($\mathbf{q}^0, \beta^0, U, P^0, K$)
2:     $\Lambda \leftarrow$ CONSTRAINTDIRECTIONS($\mathbf{q}^0, \beta^0, U$)
3:     $\Upsilon \leftarrow$ BUILDTEMPLATE($\Lambda, U, f$)
4:     $\Phi \leftarrow$ BERNCOEFFS($s \circ f \circ \gamma_G$)
5:     $j \leftarrow 1$
6:     **repeat**
7:         $P^j \leftarrow$ REFPARAMS($\mathbf{q}^{j-1}, \beta^{j-1}, P^{j-1}, \Phi$)
8:         **if** $P^j \neq \emptyset$ **then**
9:             $[\mathbf{q}^j, \beta^j] \leftarrow$ REACHSTEP($\mathbf{q}^{j-1}, \beta^{j-1}, P^j, \Upsilon$)
10:        **end if**
11:        $j \leftarrow j + 1$
12:    **until** $(j = K) \vee (P^j = \emptyset)$
13:    **return** $(j, \mathbf{q}^j, \beta^j, P^j)$
14: **end function**

---

## 5   Experimental Results

Our case study focuses on epidemiological models. Such systems are useful to understand the dynamics of infectious diseases and to plan strategies that counter their proliferation. The recent emergence of the influenza strain A(H1NA) in the United States and Mexico [15], or the Severe Acute Respiratory Syndrome (SARS) in southern China [16], are some examples that show the impact of the spread of diseases on our society. Often, vaccines are not available for the entire population, because of disproportional demand or elevated costs that many countries cannot afford. Therefore, a good alternative strategy based on faithful mathematical models can bring benefits to both the population health and country economics.

## 5.1  SIR and SARS

Let us consider the basic epidemic SIR model [17] in its continuous and Euler discretized (with time step $h$) versions:

$$
\begin{aligned}
\dot{S} &= -\beta SI & S_{k+1} &= S_k - (\beta S_k I_k)h \\
\dot{I} &= \beta SI - \gamma I & I_{k+1} &= I_k + (\beta S_k I_k - \gamma I_k)h \\
\dot{R} &= \gamma I & R_{k+1} &= R_k + (\gamma I)h
\end{aligned}
$$

In this model, a fixed population $N = S(t) + I(t) + R(t)$ is grouped in three classes: $S(t)$ is the number of individuals not yet infected and susceptible to the disease, $I(t)$ are the individuals who have been infected and who could infect healthy individuals, $R(t)$ are those who have been infected and removed from the system. The parameters $\beta$ and $1/\gamma$ are the contraction rate of the disease and the mean infective period, respectively.

We now perform two tests choosing different generator sets but with the same contraction rate $\beta = 0.34$, initial mean infective period coefficient $\gamma \in [0.05, 0.07]$, time step $h = 1.0$, and maximum number of reachability steps $K = 30$. In all the tests the population is normalized.

For the first test we fix as safety condition the constraint $s(I) = I - 0.64$, that corresponds asking whether or not there are values of $\gamma$ such that the maximum number of infected individuals stays always below 0.64. We consider a generator set whose vectors are mutually perpendicular and that leads to a box. The generator versor set $U = \{u_1, u_2, u_3\}$, template matrix $\Lambda$, and vector amplitudes $\beta$ are:

$$
\begin{array}{ccc}
\begin{aligned}
u_1 &= (1,0,0) \\
u_2 &= (0,1,0) \\
u_3 &= (0,0,1)
\end{aligned}
&
\Lambda = \begin{pmatrix}
1 & 0 & 0 \\
0 & 1 & 0 \\
0 & 0 & 1 \\
-1 & 0 & 0 \\
0 & -1 & 0 \\
0 & 0 & -1
\end{pmatrix}
&
\beta = \begin{pmatrix}
0.001 \\
0.001 \\
0.001
\end{pmatrix}.
\end{array}
$$

In the second test we strengthen the safety condition down to $s(I) = I - 0.62$. In such a case we try to keep the number of infected individuals always lower than 0.62. Choosing the versor set $U = \{u_1, u_2, u_3\}$, template matrix $\Lambda$, and vector amplitudes $\beta$ with values

$$u_1 = (0.7071, 0.7071, 0)$$
$$u_2 = (-0.7071, 0.7071, 0)$$
$$u_3 = (0, 0, 1)$$

$$\Lambda = \begin{pmatrix} 0.7071 & 0.7071 & 0 \\ -0.7071 & 0.7071 & 0 \\ 0 & 0 & 1 \\ -0.7071 & -0.7071 & 0 \\ 0.7071 & -0.7071 & 0 \\ 0 & 0 & -1 \end{pmatrix}$$

$$\beta = \begin{pmatrix} 0.0014 \\ 0.0014 \\ 0.0010 \end{pmatrix}$$

we obtain a parallelotope with only two faces parallel to the axis $R$ (those generated by $u_3$). All the other are not parallel to any axis. In both the tests the initial sets are anchored to the base vertex $\mathbf{q} = (0.8, 0.2, 0)$.

In the first test our tool computed the template Bernstein coefficients in 6.77 s and synthesized the safe parameter set $P_s = [0.661, 0.675]$ in 27.95 s against the 45.83 s of our previous technique. The second test required 13.59 s to compute the template control points and 39.70 s to synthesized the safe parameter set $P_s = [0.670, 0.675]$.

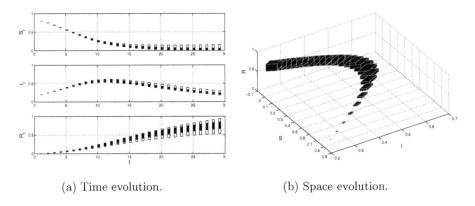

(a) Time evolution.                    (b) Space evolution.

**Fig. 1.** SIR synthesis with parallelotopes. Black and white boxes represent the constrained and unconstrained evolution of the system, respectively. Note how the edges of the parallelotopes are not parallel to the axis $S$ and $I$.

Figure 1 depicts the results of the second test. Figure 1a shows the time evolution of the system and Fig. 1b its reachability set.

In the spirit of verifying the improvements with respect to our previous approach, we now take into account a more complex and realistic epidemic model that describes the *Severe Acute Respiratory Syndrome*, commonly known as SARS [18]. The dynamics of this model are nonlinear and involve six variables and fifteen parameters. For a detailed description of the model the reader may refer to [1]. As in the previous comparison, we choose a generator versor set $U = \{u_1, u_2, \ldots, u_6\}$, template matrix $\Lambda$ and generator amplitudes $\beta$ that produce a box, and we fix the base vertex $\mathbf{q} = (6.5, 124.0, 0.0, 1.0, 0.0, 0.0)$. The simulation parameter values are the same as in [1], that is we try to synthesis the four parameters $\gamma_1, \gamma_2, k_1$, and $k_2$ imposing the safety constraint $s(I) = I - 300$.

The precomputation of the Bernstein coefficients took 14.45 s, while the parameter synthesis 907.54 s against the 2012.82 s of our previous technique. Note that for both the SIR and SARS models we have significantly reduced the computational times.

## 5.2   Influenza

In this section we consider a simplification of the influenza model described in [19]. This model is a variation of the standard SIR model where two controllable parameters, the antiviral treatment $\tau$ and the social distancing $d$, i.e., the infected individuals who receive the antiviral treatment and the number of contacts per unit time between individuals, are taken into account. The considered population is composed by $N$ individuals grouped in four classes: $S$ is the number of individuals susceptible to the influenza and not infected, $I$ are the individuals infected by the disease, $T$ are those who are under treatment, and $R$ are the recovered patients. The model is defined by the following system of difference equations:

$$
\begin{aligned}
S_{t+1} &= S_t(1 - G_t) \\
I_{t+1} &= (1 - \tau)(1 - \sigma_1)(1 - \delta)I_t + S_t G_t \\
T_{t+1} &= (1 - \sigma_2)T_t + \tau(1 - \sigma_1)(1 - \delta)I_t \\
R_{t+1} &= R_t + \sigma_1(1 - \delta)I_t + \sigma_2 T_t
\end{aligned}
$$

where $G_t = \rho(1 - d)(I_t + \varepsilon T_t)/(N_t)$. Variable $G_t$ represents the number of susceptible people that at time $t$ remains so also at time $t + 1$. The dynamics of the model involve seven parameters: $\tau$ characterizes the fraction of individuals who get the treatment; $\sigma_1$ and $\sigma_2$ are the probabilities of recovering individuals thanks to natural causes and treatment, respectively; $\delta$ is the ratio of induced deaths while $\beta$ is the disease transmission rate; $d$ is the social distancing, that is the number of contacts between individuals by unit time, and $\rho$ is the reduction in transmissibility for the treated compartment. The interesting controllable parameters are the antiviral treatment $\tau$ and the social distancing $d$.

We now simulate and study the model trying to synthesize the two controllable parameters. The recovering probabilities without and with treatment are $\sigma_1 = 1/7$ and $\sigma_2 = 1/5$; the transmissibility coefficient of the treated class is $\varepsilon = 0.7$; the mortality and susceptibility rates are fixed to $\delta = 8 \times 10^{-5}$ and $\rho = 0.5$. The controllable parameters, that are the antiviral treatment $\tau$ and the social distancing $d$, can vary inside the initial sets $\tau \in [0.001, 0.002]$ and $d \in [0.005, 0.010]$ . The imposed safety constraint is $s(I) = I - 0.3964$, while the base vertex $\mathbf{q}$, the generator versos $U = \{u_1, \dots, u_4\}$, and the vector amplitudes $\beta$ that generate the initial set are

$$
\mathbf{q} = \begin{pmatrix} 0.9 \\ 0.1 \\ 0.0 \\ 0.0 \\ 0.0 \end{pmatrix}
\qquad
\begin{aligned}
u_1 &= (0.7053, 0.7053, 0.7053, 0.0) \\
u_2 &= (0.0, 0.9806, 0.1961, 0.0) \\
u_3 &= (0.0, 0.0, 1.0, 0.0) \\
u_4 &= (0.0, 0.7071, 0.0, 0.7071)
\end{aligned}
\qquad
\beta = \begin{pmatrix} 0.1418 \\ 0.5099 \\ 0.100 \\ 0.1414 \end{pmatrix} \times 10^{-3}.
$$

As maximum number of steps we fix $K = 30$. From the initial parameter set $P = [0, 0.001, 0.002] \times [0.005, 0.010]$, our tool found the safe parameter subset $P_s \subset P$ whose vertices are $(0.0011, 0.0100)$, $(0.0020, 0.0054)$, and $(0.0020, 0.0100)$. The graphical representation of $P$ and $P_s$ is depicted in Fig. 2b. Figure 2a shows the unconstrained and constrained evolution of the influenza model. The template control points and the parameter synthesis were computed in 92.24 and 304.67 s, respectively.

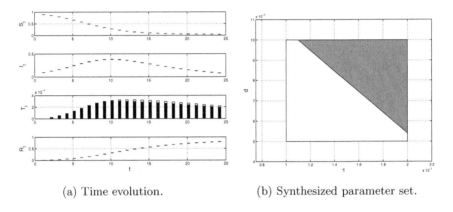

(a) Time evolution.          (b) Synthesized parameter set.

**Fig. 2.** SITR synthesis.

## 6   Conclusion

We have introduced an improved parameter synthesis algorithm for polynomial dynamical systems and shown its effectiveness by applying it to some epidemic models. An advantage of our algorithm is that it can handle a large number of parameters since their refinement can be reduced to linear program solving.

The benefits brought from the proposed advancements are twofold. First, the parallelotope based representation leads to a more precise and flexible over-approximation of the states of the considered system. Second, the introduction of the symbolical precomputation on the system dynamics and safety condition avoids the recalculation of the Bernstein coefficients at each step. We have seen how multiple evaluations of these precomputed formulæ halve the computational times with respect to our previous approach.

The results obtained from the studies on the epidemic models are encouraging and we intend to pursue this work in several directions. We will both address more complex constraints that involve logical operators and time-dependent queries and extend the set representation combining several parallelotopes hopefully obtaining better approximations and more precise parameter refinements.

# References

1. Dreossi, T., Dang, T.: Parameter synthesis for polynomial biological models. In: HSCC 2014, New York, NY, USA, pp. 233–242. ACM (2014)
2. Dang, T., Testylier, R.: Reachability analysis for polynomial dynamical systems using the Bernstein expansion. Reliable Comput. **17**(2), 128–152 (2012)
3. Ben Sassi, M.A., Testylier, R., Dang, T., Girard, A.: Reachability analysis of polynomial systems using linear programming relaxations. In: Chakraborty, S., Mukund, M. (eds.) ATVA 2012. LNCS, vol. 7561, pp. 137–151. Springer, Heidelberg (2012)
4. Testylier, R., Dang, T.: Analysis of parametric biological models with non-linear dynamics. In: International Workshop on Hybrid Systems and Biology HSB, EPTCS, vol. 92, pp. 16–29 (2012)
5. Garloff, J., Smith, A.: Rigorous affine lower bound functions for multivariate polynomials and their use in global optimisation. In: Proceedings of the 1st International Conference on Applied Operational Research. Lecture Notes in Management Science, vol. 1, pp. 199–211 (2008)
6. Barnat, J., Brim, L., Krejci, A., Streck, A., Safranek, D., Vejnar, M., Vejpustek, T.: On parameter synthesis by parallel model checking. IEEE/ACM Trans. Comput. Biol. Bioinform. **9**(3), 693–705 (2012)
7. Jha, S.K., Langmead, C.J.: Synthesis and infeasibility analysis for stochastic models of biochemical systems using statistical model checking and abstraction refinement. Theor. Comput. Sci. **412**(21), 2162–2187 (2011)
8. Bartocci, E., Bortolussi, L., Nenzi, L.: On the robustness of temporal properties for stochastic models. In: HSB 2013, EPTCS, vol. 125 (2013)
9. Grosu, R., Batt, G., Fenton, F.H., Glimm, J., Le Guernic, C., Smolka, S.A., Bartocci, E.: From cardiac cells to genetic regulatory networks. In: Gopalakrishnan, G., Qadeer, S. (eds.) CAV 2011. LNCS, vol. 6806, pp. 396–411. Springer, Heidelberg (2011)
10. Dreossi, T., Dang, T.: Falsifying oscillation properties of parametric biological models. In: HSB 2013, EPTCS, vol. 125, pp. 53–67 (2013)
11. Donzé, A.: Breach, a toolbox for verification and parameter synthesis of hybrid systems. In: Touili, T., Cook, B., Jackson, P. (eds.) CAV 2010. LNCS, vol. 6174, pp. 167–170. Springer, Heidelberg (2010)
12. Donzé, A., Clermont, G., Langmead, C.J.: Parameter synthesis in nonlinear dynamical systems: application to systems biology. J. Comput. Biol. **17**(3), 325–336 (2010)
13. Garloff, J., Smith, A.P.: A comparison of methods for the computation of affine lower bound functions for polynomials. In: Jermann, C., Neumaier, A., Sam, D. (eds.) COCOS 2003. LNCS, vol. 3478, pp. 71–85. Springer, Heidelberg (2005)
14. Sankaranarayanan, S., Sipma, H.B., Manna, Z.: Scalable analysis of linear systems using mathematical programming. In: Cousot, R. (ed.) VMCAI 2005. LNCS, vol. 3385, pp. 25–41. Springer, Heidelberg (2005)
15. Organization, W.H.: Pandemic (h1n1) 2009 - update 103, June 2010. http://www.who.int/csr/don/2010_06_04/en/
16. Organization, W.H.: Summary of probable sars cases with onset of illness from 1 November 2002 to 31 July 2003, December 2013. http://www.who.int/csr/sars/country/table2004_04_21/en/
17. Kermack, W., McKendrick, A.: A contribution to the mathematical theory of epidemics. Proc. Roy. Soc. Lond.: Ser. A, Phys. Math. Sci. **115**, 700–721 (1927)

18. Gumel, A., Ruan, S., Day, T., Watmough, J., Brauer, F., Driessche, V., Gabrielson, D., Bowman, C., Alexander, M., Ardal, S., et al.: Modelling strategies for controlling sars outbreaks. Proc. Roy. Soc. Lond. Ser. B: Biol. Sci. **271**(1554), 2223–2232 (2004)
19. Parra, P.A.G., Lee, S., Velzquez, L., Castillo-Chavez, C.: A note on the use of optimal control on a discrete time model of influenza dynamics. Math. Biosci. Eng. **8**(1), 183–197 (2011)

# Optimal Observation Time Points in Stochastic Chemical Kinetics

Charalampos Kyriakopoulos[(✉)] and Verena Wolf

Saarland University, Saarbrücken, Germany
**charis@cs.uni-saarland.de**

**Abstract.** Wet-lab experiments, in which the dynamics within living cells are observed, are usually costly and time consuming. This is particularly true if single-cell measurements are obtained using experimental techniques such as flow-cytometry or fluorescence microscopy. It is therefore important to optimize experiments with respect to the information they provide about the system. In this paper we make a priori predictions of the amount of information that can be obtained from measurements. We focus on the case where the measurements are made to estimate parameters of a stochastic model of the underlying biochemical reactions. We propose a numerical scheme to approximate the Fisher information of future experiments at different observation time points and determine optimal observation time points. To illustrate the usefulness of our approach, we apply our method to two interesting case studies.

## 1 Introduction

The successful calibration of mathematical models of biological processes is usually achieved by the interplay between computer simulations and wet-lab experiments. While both approaches are typically very time consuming, wet-lab experiments are costly compared to computer simulations, which is particularly true for single-molecule techniques such as flow cytometry or fluorescence microscopy. To keep the effort and costs as low as possible and to employ wet-lab resources to have maximal gain, it is possible to run computer simulations before measurements have been done in order to maximise the amount of information provided by the measurements. If the plan is to use the measurements for the estimation of unknown model parameters, then it is common practice to approach the optimal experimental design problem by considering the Fisher information. The Fisher information provides an approximation of the accuracy of parameter estimates and if the information is maximal so is the (approximated) accuracy of the estimators. While classical parameter estimation techniques compute the Fisher information of parameter values estimated based on certain observations of the system, it is also possible to do this computation before any observations are made. Thus, it is possible to compare the amount of information that a certain hypothetical experiment provides w.r.t. an unknown parameter.

© Springer International Publishing Switzerland 2015
O. Maler et al. (Eds.): HSB 2013 and 2014, LNBI 7699, pp. 83–96, 2015.
DOI: 10.1007/978-3-319-27656-4_5

Here, we focus on stochastic models of biochemical reaction networks, for which the optimal experimental design problem has been addressed rarely in the past. We assume that the kinetic constants of the chemical reactions have to be estimated and that flow cytometry measurements are possible. Our goal is to find the optimal times at which measurements should be made to maximize the amount of information provided by the observations. Note that such observations are not correlated over time, thus, the results are independent single-cell measurements. We do not optimize over other experiment design criteria such as the choice between the chemical species to observe. The reason is that the latter problem is, compared to the optimal observation time point problem, much simpler since we only have to compare the Fisher information for all possible combinations of observed species. Finding optimal observation times, however, is a challenging problem since the computation of the Fisher information relies on a transient solution of the model. It is therefore very costly and can usually not be done for a very large number of time points. Thus, a sophisticated numerical procedure is necessary to efficiently determine optimal observation times. Another problem is that the Fisher information depends not only on the time points the observations are made, but also on the unknown parameters value. We therefore assume that prior knowledge about the unknown parameters is available in the form of a prior distribution. Given such a prior and a stochastic model with unknown kinetic constants, we determine those observation time points at which the expected Fisher information is maximal.

There are two previous approaches to the optimal experimental design problem for stochastic chemical kinetics. Both of them provide approximations of the Fisher information and focus on systems for which a direct numerical computation of the transient solution is too expensive. Komorowski et al. propose an approach that is based on the linear noise approximation, which assumes that molecules are present in sufficiently high copy numbers [7]. However, many systems involve species present in small copy numbers leading to highly skewed distributions [15]. In such cases the linear noise approximation yields poor approximations and the underlying master equation has to be solved directly. Ruess et al. propose an approximation of the Fisher information based on the moments of the underlying probability distribution. Assuming that the number of observed cells is large, they derive an expression for the Fisher information that only involves moments up to order four [15]. This derivation is based on the fact that sample mean and sample variance of the observations are approximately normally distributed. However, it turns out that the information provided by the sample mean and the sample variance do not suffice for the characterization of skewed or bimodal distributions. In this paper we use a method based on the direct approximation of the underlying probability distribution since we found approximation errors up to 40 % when using the approach of Ruess et al. We combine this approximation with a gradient-descent based optimization scheme in which we, after sampling from the prior distribution, have to solve the underlying master equation only once over time. The latter is achieved by a direct numerical approximation where the large state space of the model is truncated dynamically. This approach is a modification of an earlier method developed for the estimation of parameters [1].

After introducing the stochastic model in Sect. 2, we give the mathematical background for the estimation of unknown parameters and explain how the Fisher information and its time-derivatives are computed in Sect. 3. In Sect. 4 we present the optimal experiment design problem and propose a numerical method for finding optimal observation times. Finally, we report on experimental results for two reaction networks (Sect. 5) and conclude with a discussion of the results in Sect. 6.

## 2   Discrete-State Stochastic Model

According to Gillespie's theory of stochastic chemical kinetics, a well-stirred mixture of $n$ molecular species in a volume with fixed size and fixed temperature can be represented as a continuous-time Markov chain $\{\mathbf{X}_t, t \geq 0\}$ [5]. The random vector $\mathbf{X}_t$ describes the chemical populations at time $t$, i.e., the $i$-th entry is the number of molecules of type $i \in \{1, \ldots, n\}$ at time $t$. Thus, the state space of $\mathbf{X}$ is $\mathbb{Z}_+^n = \{0, 1, \ldots\}^n$. The state changes of $\mathbf{X}$ are triggered by the occurrences of chemical reactions, which are of $m$ different types. For $j \in \{1, \ldots, m\}$ let the row vector $\mathbf{v}_j \in \mathbb{Z}^n$ be the nonzero *change vector* of the $j$-th reaction type, that is, $\mathbf{v}_j = \mathbf{v}_j^- + \mathbf{v}_j^+$ where $\mathbf{v}_j^-$ contains only non-positive entries, which specify how many molecules of each species are consumed (*reactants*) if an instance of the reaction occurs. The vector $\mathbf{v}_j^+$ contains only non-negative entries, which specify how many molecules of each species are produced (*products*). Thus, if $\mathbf{X}_t = \mathbf{x}$ for some $\mathbf{x} \in \mathbb{Z}_+^n$ with $\mathbf{x} + \mathbf{v}_j^-$ being non-negative, then $\mathbf{X}_{t+dt} = \mathbf{x} + \mathbf{v}_j$ is the state of the system after the occurrence of the $j$-th reaction within the infinitesimal time interval $[t, t + dt)$.

Each reaction type has an associated *propensity function*, denoted by $\alpha_1, \ldots, \alpha_m$, which is such that $\alpha_j(\mathbf{x}) \cdot dt$ is the probability that, given $\mathbf{X}_t = \mathbf{x}$, one instance of the $j$-th reaction occurs within $[t, t + dt)$. Often the value $\alpha_j(\mathbf{x})$ is chosen proportional to the number of distinct reactant combinations in state $\mathbf{x}$, known as the law of mass action. However, for many reactions the proportionality constant $\theta_j$ is unknown and has to be estimated based on measurements. For instance, if we have two (distinct) reactants (i.e. $\mathbf{v}_j^- = -\mathbf{e}_i - \mathbf{e}_\ell$) then $\alpha_j(\mathbf{x}) = \theta_j \cdot x_i \cdot x_\ell$ where $x_i$ and $x_\ell$ are the corresponding entries of $\mathbf{x}$, $i \neq \ell$, $\theta_j > 0$, and $\mathbf{e}_i$ is the vector with the $i$-th entry 1 and all other entries 0. In the sequel we do not restrict the form of $\alpha_j$ but only assume that its derivative w.r.t. some unknown parameter $\theta_j$ exists. Sometimes we will make the dependence of $\alpha_j$ on $\theta_j$ explicit by writing $\alpha_j(\mathbf{x}, \theta_j)$ instead of $\alpha_j(\mathbf{x})$.

*Example 1.* We consider a simple crystallization process that involves four chemical species, namely A, B, C and D. Thus, the entries of the random vector $\mathbf{X}_t$ give the numbers of molecules of types A, B, C and D at time $t$. The two possible reactions are 2A $\rightarrow$ B and A + C $\rightarrow$ D. Thus, $\mathbf{v}_1 = (-2, 1, 0, 0)$, $\mathbf{v}_2 = (-1, 0, -1, 1)$. For a state $\mathbf{x} = (x_A, x_B, x_C, x_D)$, the propensity functions are $\alpha_1(\mathbf{x}) = \theta_1 \cdot \binom{x_A}{2}$ and $\alpha_2(\mathbf{x}) = \theta_2 \cdot x_A \cdot x_C$. Note that given an initial state $\mathbf{x}_0$ the set of reachable states is a finite subset of $\mathbb{N}^4$.

In general, the reaction rate constants $\theta_j$ refer to the probability that a randomly selected pair of reactants collides and undergoes the $j$-th chemical reaction. It depends on the volume and the temperature of the system as well as on the microphysical properties of the reactant species. Since reactions of higher order (requiring more than two reactants) are usually the result of several successive lower order reactions, we do not consider the case of more than two reactants.

**The Chemical Master Equation.** For $\mathbf{x} \in \mathbb{Z}_+^n$ and $t \geq 0$, let $p_t(\mathbf{x})$ denote the probability $Pr(\mathbf{X}_t = \mathbf{x})$. Given $\mathbf{v}_1^-, \ldots, \mathbf{v}_m^-, \mathbf{v}_1^+, \ldots, \mathbf{v}_m^+, \alpha_1, \ldots, \alpha_m$, and some initial distribution $\mathbf{p}_0$, the Markov chain $\mathbf{X}$ is uniquely specified and its evolution is given by the chemical master equation (CME)

$$\tfrac{d}{dt} p_t(\mathbf{x}) = \sum_{j:\mathbf{x}-\mathbf{v}_j^- \geq 0} \alpha_j(\mathbf{x} - \mathbf{v}_j) p_t(\mathbf{x} - \mathbf{v}_j) - \alpha_j(\mathbf{x}) p_t(\mathbf{x}). \tag{1}$$

If we use $\mathbf{p}_t$ to denote the row vector with entries $p_t(\mathbf{x})$, then the vector form of the CME becomes

$$\tfrac{d}{dt} \mathbf{p}_t = \mathbf{p}_t Q, \tag{2}$$

where $Q$ is the infinitesimal generator matrix of $\mathbf{X}$ with $Q(\mathbf{x}, \mathbf{y}) = \alpha_j(\mathbf{x})$ if $\mathbf{y} = \mathbf{x} + \mathbf{v}_j$ and $\mathbf{x} + \mathbf{v}_j^- \geq 0$. Note that, in order to simplify our presentation, we assume here that all vectors $\mathbf{v}_j$ are distinct. All remaining entries of $Q$ are zero except for the diagonal entries which are equal to the negative row sum. The ordinary first-order differential equation in (2) is a direct consequence of the Kolmogorov forward equation. Since $\mathbf{X}$ is a regular Markov process, (2) has the general solution $\mathbf{p}_t = \mathbf{p}_0 \cdot e^{Qt}$, where $e^A$ is the matrix exponential of a matrix $A$. If the state space of $X$ is infinite, then we can only compute approximations of $\mathbf{p}_t$. But even if $Q$ is finite, its size is often large because it grows exponentially with the number of state variables. Therefore standard numerical solution techniques for systems of first-order linear equations of the form of (2) are infeasible. The reason is that the number of nonzero entries in $Q$ often exceeds the available memory capacity for systems of realistic size. If the populations of all species remain small (at most a few hundreds) then the CME can be efficiently approximated using projection methods [4,6,12] or fast uniformization methods [10,16]. The idea of these methods is to avoid an exhaustive state space exploration and, depending on a certain time interval, restrict the analysis of the system to a subset of states.

Here, we are also interested in the partial derivatives of $\mathbf{p}_t$ w.r.t. the reaction rate constants $\theta = (\theta_1, \ldots, \theta_m)$. In the sequel we will write $\mathbf{p}_t(\theta)$ instead of $\mathbf{p}_t$ to make the dependency on $\theta$ explicit and the entry of $\mathbf{p}_t(\theta)$ that corresponds to state $\mathbf{x}$ will be denoted by $p_t(\mathbf{x}; \theta)$. Moreover, we define the row vectors $\mathbf{s}_t^j(\theta)$ as the derivative of $\mathbf{p}_t(\theta)$ w.r.t. $\theta_j$, i.e.,

$$\mathbf{s}_t^j(\theta) = \tfrac{\partial \mathbf{p}_t(\theta)}{\partial \theta_j} = \lim_{\Delta h \to 0} \tfrac{\mathbf{p}_t(\theta + \Delta \mathbf{h}^{(j)}) - \mathbf{p}_t(\theta)}{\Delta h},$$

where the vector $\mathbf{\Delta}h^{(j)}$ is zero everywhere except for the $j$-th position that is equal to $\Delta h$. We denote the entry in $\mathbf{s}_t^j(\theta)$ that corresponds to state $\mathbf{x}$ by $s_t^j(\mathbf{x}, \theta)$. Derivating (2), we find that $\mathbf{s}_t^j(\theta)$ is the unique solution of the following linear system of ODEs

$$\tfrac{d}{dt}\mathbf{s}_t^j(\theta) = \mathbf{s}_t^j(\theta)Q + \mathbf{p}_t(\theta)\tfrac{\partial}{\partial\theta_j}Q, \tag{3}$$

where $j \in \{1, \ldots, m\}$. The initial condition is $s_0^j(\mathbf{x}, \theta) = 0$ for all $\mathbf{x}$ and $\theta$ since $p_0(\mathbf{x}; \theta)$ is independent of $\theta_j$.

# 3 Observations and Fisher Information

Following the notation in [14], we assume that observations of a biochemical network are made at time instances $t_1, \ldots, t_R \in \mathbb{R}_{\geq 0}$, where $t_1 \leq \ldots \leq t_R$. The entries of the random vector $\mathbf{O}_{t_k} \in \mathbb{R}^n$ describe the molecule numbers observed at time $t_k$ for $k \in \{1, \ldots, R\}$. Since these observations are typically subject to measurement errors, we may assume that $\mathbf{O}_{t_k} = \mathbf{X}_{t_k} + \epsilon_{t_k}$, where the entries of the error terms $\epsilon_{t_k}$ are independent and identically normally distributed with mean zero and standard deviation $\sigma$. Note that $\mathbf{X}_{t_k}$ is the true population vector at time $t_k$. Clearly, this implies that, conditional on $\mathbf{X}_{t_k}$, the random vector $\mathbf{O}_{t_k}$ is independent of all other observations as well as independent of the history of $\mathbf{X}$ before time $t_k$.

Let $f$ denote the joint density of $\mathbf{O}_{t_1}, \ldots, \mathbf{O}_{t_R}$ and assume that $\mathbf{o}_1, \ldots, \mathbf{o}_R \in \mathbb{R}^n$. Then the likelihood of the observation sequence $\mathbf{o}_{t_1}, \ldots, \mathbf{o}_{t_R}$ is

$$\begin{aligned}
\mathcal{L} &= f\left(\mathbf{O}_{t_1} = \mathbf{o}_1, \ldots, \mathbf{O}_{t_R} = \mathbf{o}_R\right) \\
&= \sum_{\mathbf{x}_1} \cdots \sum_{\mathbf{x}_R} f\left(\mathbf{O}_{t_1} = \mathbf{o}_1, \ldots, \mathbf{O}_{t_R} = \mathbf{o}_R \mid \mathbf{X}_{t_1} = \mathbf{x}_1, \ldots, \mathbf{X}_{t_R} = \mathbf{x}_R\right) \quad (4) \\
&\quad Pr(\mathbf{X}_{t_1} = \mathbf{x}_1, \ldots, \mathbf{X}_{t_R} = \mathbf{x}_R).
\end{aligned}$$

We assume that for the unobserved process $\mathbf{X}$ we do not know the values of the rate constants $\theta = (\theta_1, \ldots, \theta_m)$ and our aim is to estimate these constants.

In the sequel, in order to keep the notation simpler, we assume no measurement errors in our observations. Nevertheless, it is straightforward to extend our proposed optimal design procedure to the case where the measurements are not exact. As shown in [1] this only introduces additional weights during the calculation of the likelihood. In this case, also, one can consider the standard deviation of the error terms, $\sigma$, as an unknown parameter and apply the proposed design. Here, though, we assume that $\mathbf{O}_{t_k} = \mathbf{X}_{t_k}$ for all $k$ and for a concrete observation sequence $\mathbf{x}_1, \ldots, \mathbf{x}_R \in \mathbb{N}^n$ the likelihood becomes

$$\mathcal{L} = Pr(\mathbf{X}_{t_1} = \mathbf{x}_1, \ldots, \mathbf{X}_{t_R} = \mathbf{x}_R). \tag{5}$$

Of course, $\mathcal{L}$ depends on the chosen rate parameters $\theta$ since the probability measure $Pr(\cdot)$ does. When necessary, we will make this dependence explicit by writing $\mathcal{L}(\theta)$ instead of $\mathcal{L}$. Since our observations are equal to the true state $\mathbf{X}_{t_k}$

at time $t_k$, we write $\mathcal{L}(\mathbf{X}; \theta)$ where with some abuse of notation $\mathbf{X}$ now refers to the sequence $\mathbf{X}_{t_1}, \ldots, \mathbf{X}_{t_R}$. We now seek constants $\theta^*$ such that

$$\theta^* = \operatorname{argmax}_\theta \mathcal{L}(\mathbf{X}; \theta), \tag{6}$$

where the maximum is taken over all vectors $\theta$ with all components strictly positive. This optimization problem is known as the maximum likelihood problem [8]. Note that $\theta^*$ is a random variable, in the sense it depends on the (random) observations $\mathbf{X} = (\mathbf{X}_{t_1}, \ldots, \mathbf{X}_{t_R})$.

The maximum likelihood estimator $\theta^*$ is known to be asymptotically normally distributed and its covariance matrix approaches the Cramér-Rao bound $\mathcal{I}_\mathbf{X}(\theta)^{-1}$, where $\mathcal{I}_\mathbf{X}(\theta)$ is the Fisher information matrix (FIM). Note that this bound is a lower bound on the estimator's covariance matrix. It is commonly used to derive confidence intervals for the estimated values of the parameters. Previous experimental results show that the variances approximated based on the FIM are close to the variances approximated based on many repetitions of the experiments and the estimation procedure [1]. Thus, in order to have an accurate estimation of the parameters the (co-)variances should be as small as possible, i.e., $\mathcal{I}_\mathbf{X}(\theta)$ should be large to achieve tight confidence intervals for $\theta^*$.

Given an observation sequence, $\mathbf{X}$ of the process, the entry of the FIM that corresponds to the unknown parameters $\theta_i$ and $\theta_j$, $1 \leq i, j \leq m$ is defined as

$$\left(\mathcal{I}_\mathbf{X}(\theta)\right)_{i,j} = \mathbb{E}_\mathbf{X}\left[\left(\tfrac{\partial}{\partial \theta_i}\log \mathcal{L}(\mathbf{X}; \theta)\right)\left(\tfrac{\partial}{\partial \theta_j}\log \mathcal{L}(\mathbf{X}; \theta)\right)\right]. \tag{7}$$

Note that the expectation is taken w.r.t. the observation sequence $\mathbf{X} = (\mathbf{X}_{t_1}, \ldots, \mathbf{X}_{t_R})$. Under certain (mild) regularity conditions that hold for the likelihoods we consider here, Eq. (7) can be equivalently written as [17]

$$\begin{aligned}
\left(\mathcal{I}_\mathbf{X}(\theta)\right)_{i,j} &= -\mathbb{E}_\mathbf{X}\left[\tfrac{\partial^2}{\partial \theta_i \partial \theta_j}\log \mathcal{L}(\mathbf{X}; \theta)\right] \\
&= -\sum_{\mathbf{x}_1, \ldots, \mathbf{x}_R} Pr(\mathbf{X}_{t_1} = \mathbf{x}_1, \ldots, \mathbf{X}_{t_R} = \mathbf{x}_R; \theta) \\
&\qquad \tfrac{\partial^2}{\partial \theta_i \partial \theta_j}\log Pr(\mathbf{X}_{t_1} = \mathbf{x}_1, \ldots, \mathbf{X}_{t_R} = \mathbf{x}_R; \theta).
\end{aligned} \tag{8}$$

Note that if the observations $\mathbf{X}_{t_1}, \ldots, \mathbf{X}_{t_R}$ are independent observations then the $(i, j)$th entry of the FIM is such that

$$\left(\mathcal{I}_\mathbf{X}(\theta)\right)_{i,j} = \sum_{k=1}^{R} \left(\mathcal{I}_{\mathbf{X}_{t_k}}(\theta)\right)_{i,j} \tag{9}$$

where $\mathcal{I}_{\mathbf{X}_{t_k}}(\theta)$ is the Fisher information matrix of a single observation $\mathbf{X}_{t_k}$ at time $t_k$. The above means that the information of a sequence of $R$ independent observations is simply the sum of the information of each. This makes the computation of the total information easier than in the dependent case since it is enough to solve the CME along with the partial derivatives $\mathbf{s}_t^j(\theta)$, for all $j$, until time point $t_R$. This can be easily seen by exploiting (7) for $\mathbf{X}_{t_k}$.

$$\left(\mathcal{I}_{\mathbf{X}_{t_k}}(\theta)\right)_{i,j} = \mathbb{E}_{\mathbf{X}_{t_k}}\left[\left(\frac{\partial}{\partial\theta_i}\log Pr(\mathbf{X}_{t_k};\theta)\right)\left(\frac{\partial}{\partial\theta_j}\log Pr(\mathbf{X}_{t_k};\theta)\right)\right]$$

$$= \sum_{\mathbf{x}_k}\left(\frac{\partial}{\partial\theta_i}\log p_{t_k}(\mathbf{x}_k;\theta)\right)\left(\frac{\partial}{\partial\theta_j}\log p_{t_k}(\mathbf{x}_k;\theta)\right)p_{t_k}(\mathbf{x}_k;\theta) \tag{10}$$

$$= \sum_{\mathbf{x}_k}\frac{1}{p_{t_k}(\mathbf{x}_k;\theta)}\frac{\partial}{\partial\theta_i}p_{t_k}(\mathbf{x}_k;\theta)\frac{\partial}{\partial\theta_j}p_{t_k}(\mathbf{x}_k;\theta),$$

where the sums run over all possible states $\mathbf{x}_k$ that can be observed at time $t_k$. Using the notation of the previous section, we get

$$\left(\mathcal{I}_{\mathbf{X}_{t_k}}(\theta)\right)_{i,j} = \sum_{\mathbf{x}_k}\frac{s_{t_k}^i(\mathbf{x}_k;\theta)\,s_{t_k}^j(\mathbf{x}_k;\theta)}{p_{t_k}(\mathbf{x}_k;\theta)}. \tag{11}$$

Derivating Eq. (11) we can also compute the derivative of the FIM w.r.t. the time point $t_k$.

$$\frac{\partial}{\partial t}\left(\mathcal{I}_{\mathbf{X}_{t_k}}(\theta)\right)_{i,j} = \sum_{\mathbf{x}_k}\frac{\frac{\partial}{\partial t}s_{t_k}^i(\mathbf{x}_k;\theta)\,s_{t_k}^j(\mathbf{x}_k;\theta) + s_{t_k}^i(\mathbf{x}_k;\theta)\frac{\partial}{\partial t}s_{t_k}^j(\mathbf{x}_k;\theta)}{p_{t_k}(\mathbf{x}_k;\theta)}$$

$$- \sum_{\mathbf{x}_k}\frac{s_{t_k}^i(\mathbf{x}_k;\theta)\,s_{t_k}^j(\mathbf{x}_k;\theta)\frac{\partial}{\partial t}p_{t_k}(\mathbf{x}_k;\theta)}{p_{t_k}(\mathbf{x}_k;\theta)^2}. \tag{12}$$

The time derivative of the FIM is particularly useful for an efficient gradient-based optimization scheme to find the time points that provide the maximum information which is the main goal of the next section.

## 4 Optimal Observation Time Points

Assume now that an experiment is planned where the system is observed at time points $t_1,\dots,t_R$ and that the observations will be independent (e.g. if the chosen measurement technique is flow cytometry). We want to find the optimal times points to take our observations, i.e. those which yield the maximum information for the system parameters $\theta$. However, the intrinsic problem in experiment design is that the parameter values $\theta$ are, of course, unknown before the experiment is set up. Here, the chosen approach to overcome this obstacle is to search for the observation time points that maximize the determinant of the *expected* FIM when one assumes a prior distribution for the unknown parameters as suggested, for instance, in [11]. In other words our goal is to find

$$\mathbf{t}^* = \underset{\mathbf{t}=(t_1,\dots,t_R)}{\mathrm{argmax}}\ \det\left(\mathbb{E}_\theta[\mathcal{I}(\mathbf{t},\theta)]\right). \tag{13}$$

This criterion is known to be robust because it incorporates a prior belief for $\theta$. Note that in the sequel we write $\mathcal{I}(\mathbf{t},\theta)$ instead of $\mathcal{I}_{\mathbf{X}}(\theta)$ to explicitly indicate that the information is a function of the sequence of time points, $\mathbf{t}$, at which the observations are to be taken. Also, remember that in the previous section we showed that $\mathcal{I}(\mathbf{t}, \theta) = \sum_{k=1}^{R} \mathcal{I}(t_k, \theta)$, when we assume that the observations are independent.

It is worth mentioning that sometimes not all reaction rate constants are of interest, in which case we partition the parameter vector $\theta = [\theta_{(1)}, \theta_{(2)}]$ in parameters of interest $\theta_{(1)}$ and nuisance parameters $\theta_{(2)}$. Then, for a specific time point $t$, $\mathcal{I}(t, \theta)$ is replaced by the matrix $\mathcal{I}_s(t, \theta)$ which is defined as

$$\mathcal{I}_s(t, \theta) = \mathcal{I}_{11}(t, \theta) - \mathcal{I}_{12}(t, \theta) \, \mathcal{I}_{22}^{-1}(t, \theta) \, \mathcal{I}_{12}^{\mathsf{T}}(t, \theta), \text{ where}$$

$$\mathcal{I}(t, \theta) = \begin{bmatrix} \mathcal{I}_{11}(t, \theta) & \mathcal{I}_{12}(t, \theta) \\ \mathcal{I}_{12}^{\mathsf{T}}(t, \theta) & \mathcal{I}_{22}^{-1}(t, \theta) \end{bmatrix}. \tag{14}$$

Here, subscript $s$ indicates that we only consider derivatives of the parameters of interest, i.e., the matrices $\mathcal{I}_{11}(t, \theta)$, $\mathcal{I}_{22}(t, \theta)$ contain information about the variance of $\theta_{(1)}$ and $\theta_{(2)}$, respectively, while $\mathcal{I}_{12}(t, \theta)$ approximates the covariance matrix of $\theta_{(1)}$ and $\theta_{(2)}$. Hence, given a prior of $\theta$ and having computed the matrices $\mathcal{I}(t_k, \theta)$ for all $\theta$ and all $k$, it is straightforward to compute the matrix $\mathbb{E}_\theta[\mathcal{I}_s(\mathbf{t}, \theta)]$ and then $\det(\mathbb{E}_\theta[\mathcal{I}_s(\mathbf{t}, \theta)])$. Similarly, if we have, in addition, $\frac{\partial}{\partial t}\mathcal{I}(t, \theta)$ for all $\theta$ and all $t \in \{t_1, \ldots, t_R\}$ we can also compute $\frac{\partial}{\partial t}\det(\mathbb{E}_\theta[\mathcal{I}_s(\mathbf{t}, \theta)])$ as follows. From Jacobi's formula it holds that for any square matrix $A$

$$\frac{\partial}{\partial t}\det(A) = \mathrm{tr}(\mathrm{adj}(A))\frac{\partial}{\partial t}A, \tag{15}$$

where $\mathrm{tr}(A)$ is defined as the sum of the elements of the main diagonal of $A$ and $\mathrm{adj}(A)$ is the adjoint matrix of $A$. Here, $A = \mathbb{E}_\theta[\mathcal{I}_s(\mathbf{t}, \theta)]$ and, derivating Eq. (14), we can compute $\frac{\partial}{\partial t}\mathbb{E}_\theta[\mathcal{I}_s(\mathbf{t}, \theta)]$ by exploiting known matrix calculus identities.

$$\begin{aligned}\frac{\partial}{\partial t}\mathbb{E}_\theta[\mathcal{I}_s(\mathbf{t}, \theta)] = \mathbb{E}_\theta[&\tfrac{\partial}{\partial t}\mathcal{I}_{11}(\mathbf{t}, \theta) - \tfrac{\partial}{\partial t}\mathcal{I}_{12}(\mathbf{t}, \theta) \, \mathcal{I}_{22}^{-1}(\mathbf{t}, \theta) \, \mathcal{I}_{12}^{\mathsf{T}}(\mathbf{t}, \theta) \\ &- \mathcal{I}_{12}(\mathbf{t}, \theta) \, \tfrac{\partial}{\partial t}\mathcal{I}_{22}^{-1}(\theta) \, \mathcal{I}_{12}^{\mathsf{T}}(\mathbf{t}, \theta) \\ &- \mathcal{I}_{12}(\mathbf{t}, \theta) \, \mathcal{I}_{22}^{-1}(\mathbf{t}, \theta) \, \tfrac{\partial}{\partial t}\mathcal{I}_{12}^{\mathsf{T}}(\mathbf{t}, \theta)],\end{aligned} \tag{16}$$

where the derivative of the inverse is computed as

$$\tfrac{\partial}{\partial t}\mathcal{I}_{22}^{-1}(\mathbf{t}, \theta) = -\mathcal{I}_{22}^{-1}(\mathbf{t}, \theta) \, \tfrac{\partial}{\partial t}\mathcal{I}_{22}(\mathbf{t}, \theta) \, \mathcal{I}_{22}^{-1}(\mathbf{t}, \theta).$$

The main computational effort in the search for the optimal experiment is that for finding $\mathbf{t}^*$ as defined in Eq. (13), we have to solve the CME after sampling from the prior of $\theta$. It is generally not possible to find $\mathbf{t}^*$ from a single solution of the CME since we need to average over all possible values for $\theta = (\theta_1, \ldots, \theta_m)$. Therefore, we propose the following gradient descent based procedure to approximate local maxima of the determinant of the expected FIM:

1. Given a prior $\mu_\theta$ for the distribution of $\theta$, sample values $\theta^{(1)}, \ldots, \theta^{(N)} \sim \mu_\theta$.
2. Choose a sequence $\mathbf{t}_{\mathrm{next}} = (t_1, \ldots, t_R)$ of time points and for each sample of $\theta^{(i)}$ compute $\mathcal{I}_s(\mathbf{t}_{\mathrm{next}}, \theta^{(i)})$ and $\frac{\partial}{\partial t}\mathcal{I}_s(\mathbf{t}_{\mathrm{next}}, \theta^{(i)})$ for $t \in \{t_1, \ldots, t_R\}$.

3. Return approximations of $\det\left(\mathbb{E}_\theta[\mathcal{I}_s(\mathbf{t},\theta)]\right)$ and $\frac{\partial}{\partial t}\det\left(\mathbb{E}_\theta[\mathcal{I}_s(\mathbf{t},\theta)]\right)$ by averaging over the results for $i = 1,\dots,N$.
4. Following the gradient choose $\mathbf{t}_{\text{next}} = (t'_1,\dots,t'_R)$ and repeat from 2 until you find a local maximum.

Clearly, an approximation of the global maximum is found by starting the local gradient based search from multiple initial points. The initial points used in the second step can be chosen randomly or according to some heuristics as it is usual for global optimization methods. A technical but computationally important detail is that there is no need to solve the CME for every sequence of time points that is considered in the optimization algorithm. We can solve the CME once and keep the values of the Fisher information matrix and its derivatives over time and recall it for every new sequence of time points (up to a chosen discretization). Consequently, the total computational effort of the above optimization procedure is to solve once the CME until time point equal to the maximal value of $t_R$ encountered during the optimization.

Certainly, the above optimality criterion is by no means the only possible choice. A slightly different approach, for instance, would be to maximize the expected determinant of FIM [13]. Alternatively, if no prior is available, it is also possible to consider $\max_{\mathbf{t}} \min_\theta \det\left(\mathcal{I}_s(\mathbf{t},\theta)\right)$ to make sure that for any choice of $\theta$ the experiment provides maximal information. Here, we assume that a prior is available because, most probably, one experiment has been already done in order to acquire some prior knowledge for the parameters. At last, an additional advantage of a robust optimal design is that in case there is the option to perform more than two experiments in total, the above procedure can be used in iterations alternating between experiments and the update of the prior of $\theta$ via parameter estimation from the real observations.

## 5   Experimental Results

We consider two biochemical reaction networks to which we apply our experiment design procedure, namely the crystallization model, described in Example 1, and the so-called exclusive switch model [9]. The crystallization model is a very simple example because it has a finite state space and the CME can be integrated directly if the initial molecule numbers are not particularly large. However, for large initial conditions a transient solution is only possible if the state space is dynamically truncated as suggested in [10] (see also [2]). The second example is infinite in two dimensions and its distribution is bimodal. Thus, for this system we solved the CME using a dynamic truncation of the state space whenever necessary. We chose the truncation threshold $e^{-20}$ and the dynamic truncation of the states are based on the ratio in Eq. (11), rather than only on the value of $p_t(\mathbf{x};\theta)$. More precisely, a state $\mathbf{x}$ is considered as significant at time $t$ whenever this ratio is greater than the truncation threshold while in [10] it is only the current probability that determines whether a state is considered or not at time $t$.

## 5.1  Crystallization

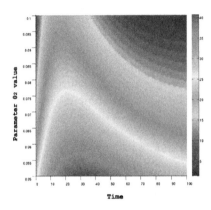

**Fig. 1.** Approximated fisher information of the crystallization example for different values of $\theta_2$ (Color figure online).

Recall the simple crystallization process given by the two chemical reactions

$$2A \xrightarrow{\theta_1} B,$$

$$A + C \xrightarrow{\theta_2} D.$$

Let us initially assume that the value of $\theta_1 = 4$ is known and we have a prior for $\theta_2 \sim \mathbb{U}(0.05, 0.1)$ where $\mathbb{U}(a, b)$ refers to the continuous uniform distribution between $a$ and $b$. For the initial state $(x_A, x_B, x_C, x_D) = (4, 0, 2, 0)$ the number of reachable states is 7 and no sophisticated truncation method is necessary to integrate the CME. Our goal is to find the best time points to take an observation for estimating $\theta_2$. In this case, the Fisher information is just a scalar and in Fig. 1 it is shown as a function of both $\theta_2$ and time. The different colors correspond to different values of $\mathcal{I}(t, \theta_2)$ (see colorbar). The time point for the maximum expected information is $t^* = 20.64$. Note that in case of one unknown parameter the extension to the $R$ time points optimization problem is straightforward because of the linearity of (9). The solution simply consists of $R$ replications of $t^*$.

Next, we assume that both, $\theta_1$ and $\theta_2$ are unknown parameters of interest. Then the Fisher information is a $2 \times 2$ matrix. In Fig. 2(a) the evolution of the determinant of the expected FIM is shown for $\theta_1 \sim \mathbb{U}(0.05, 0.5)$ and $\theta_2 \sim \mathbb{U}(0.01, 0.1)$. From the plot it is clear that if there is only one observation possible this has to be done quite early in time. Indeed, our optimal design scheme returns that the optimal time point to take an observation is at $t = 4.625$ assuming that both parameters have to be estimated.

Now, we consider the case that we are able to take a second measurement $t_2 \geq t_1$. The determinant of the average FIM is shown in Fig. 2(b). The plot suggests that one obtains the maximum information for the estimation of both parameters if both observations take place early in the experiment. Our optimization procedure returned the optimal time points $t_1 = t_2 = 4.625$. One notes that combining an early observation with a later seems also a good choice. On the other hand, taking two late observations provides significantly less information leading possibly to parameter identifiability problems.

Generalizing the above observations for the case of $R$ time points we verify from our experiments until $R = 5$ that for this particular model and this choice of parameters the optimal experiment persistently consists of taking as many as possible observations at the same early time point. We expect the same to hold for any $R \geq 2$.

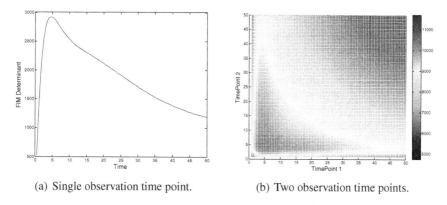

(a) Single observation time point.          (b) Two observation time points.

**Fig. 2.** Approximated determinant of the expected FIM over time for the crystallization example.

## 5.2  Exclusive Switch

The exclusive switch is a gene regulatory network that consists of two genes with a common promotor region as shown in Fig. 3. The system involves five chemical species $DNA$, $P_1$, $P_2$, $DNA.P_1$, $DNA.P_2$. At each time point the system can be in one of the following three configurations: (a) The promotor region is free, (b) $P_1$ binds to the promotor region or (c) $P_2$ binds to the promotor region. Each of the two gene products $P_1$ and $P_2$ inhibits the expression of the other product if a molecule is bound to the promotor region. More precisely, in configuration (a) (promotor region is free), molecules of both types $P_1$ and $P_2$ are produced. If a molecule of type $P_1(P_2)$ is bound to the promotor region (case (b) and (c)), only molecules of type $P_1(P_2)$ are produced, respectively. The chemical reactions with the corresponding constant rates are shown below for $j = \{1,2\}$.

$$
\begin{aligned}
DNA &\xrightarrow{\lambda_j} DNA + P_j & &\text{production} \\
P_j &\xrightarrow{\delta_j} \emptyset & &\text{degradation} \\
DNA + P_j &\xrightarrow{\beta_j} DNA.P_j & &\text{binding} \\
DNA.P_j &\xrightarrow{\nu_j} DNA + P_j & &\text{unbinding} \\
DNA.P_j &\xrightarrow{\lambda_j} DNA.P_j + P_j & &\text{bound production}
\end{aligned}
$$

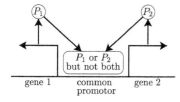

**Fig. 3.** Exclusive switch network

Depending on the chosen parameters, the probability distribution of the exclusive switch is bistable, i.e. most of the probability mass concentrates on two distinct regions of the state space. In particular, if binding to the promotor is likely, then these two regions correspond to the two configurations (b) and (c) where either the production of $P_1$ or the production of $P_2$ is inhibited.

For the purpose of our experiments we fixed the initial state of the system such that no proteins are present in the system and one DNA molecule with a free promotor region. We set up optimal experiments for the following case: The unknown parameters are the production and the degradation constants of $P_1$, $\lambda_1$ and $\delta_1$ respectively, while the rest of the parameter values are known. We assume that for $j = \{1, 2\}$

$$\lambda_1 \sim \mathbb{U}(0.01, 0.1), \delta_1 \sim \mathbb{U}(0.0001, 0.001),$$
$$\lambda_2 = 0.05, \delta_2 = 0.0005, \beta_j = 0.001 \text{ and } \nu_j = 0.008.$$

In Fig. 4(a) the information is shown over time. From the plot it is evident that in the case of a single observation time point one should take the measurement as late as possible in the interval $[0, 50]$, if we restrict until time $t = 50$. This most probably arises from the fact that the chosen binding rates, $\beta_1, \beta_2$ are rather small, i.e., binding to the promotor is not so likely and there is a delay until the binding influences the dynamics of production and degradation of the corresponding proteins.

In Fig. 4(b) the information of a possible experimental setup for two time points is being presented. From the plot we can observe that the most informative experiment, now, is given by two *different* time points. The second measurement is at 50 time units, as previously, but the first one should be taken at 16.75. Intuitively, this could mean that for estimating multiple parameters of this model we need to observe the process at more than one time points, if possible.

Setting up optimal experiments for $R$ time points we observe, as in the first model, a replication of the optimal $p$ time points, where $p$ is the number of the unknown parameters. For $R = 3$ and $R = 4$ we get $(t_1, t_2, t_3) = (16.75, 50, 50)$ and $(t_1, t_2, t_3, t_4) = (16.75, 16.75, 50, 50)$, respectively.

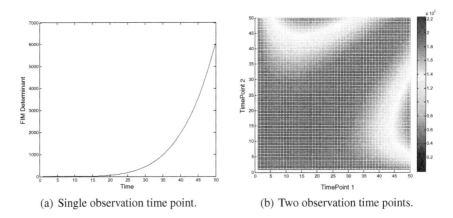

(a) Single observation time point.        (b) Two observation time points.

**Fig. 4.** Approximated determinant of the expected FIM for the exclusive switch example (case 1).

# 6    Discussion and Future Work

Given a stochastic model of a chemical reaction network, we computed the Fisher information of different experimental designs and determined optimal observation times. The optimality criterion that we considered was the determinant of the expected Fisher information where the expectation was taken w.r.t. some prior distribution over the unknown parameters.

Our experimental results give rise to the conjecture that the $n$ optimal time points for a system with $p$ unknown parameters are equal replications of the $p$ optimal time points. E.g. if we have two unknown parameters, but, say, four possible observation time points, we get only two distinct times at which the Fisher information becomes maximal w.r.t. the time points. A similar result has been proven by Box for deterministic chemical kinetics [3]. We conjecture that his result carries over to the stochastic setting. This would make experiment design for stochastic chemical kinetics a much less expensive procedure since the dimension of the optimization search space reduces to the number of unknown parameters.

Other plans for future work include that we consider equidistant observation time points and optimize the time interval between two successive observations. We also plan to consider other ways of approximating the FIM, e.g. by using moment closure techniques and exploiting the information of a sufficiently large number of moments. Additionally, we will work, also, on the case of dependent observations and develop an iterative procedure where after each optimization step experimental results become available and the next optimal observation time is computed given these results.

# References

1. Andreychenko, A., Mikeev, L., Spieler, D., Wolf, V.: Parameter identification for markov models of biochemical reactions. In: Gopalakrishnan, G., Qadeer, S. (eds.) CAV 2011. LNCS, vol. 6806, pp. 83–98. Springer, Heidelberg (2011)
2. Andreychenko, A., Mikeev, L., Spieler, D., Wolf, V.: Approximate maximum likelihood estimation for stochastic chemical kinetics. EURASIP J. Bioinf. Syst. Biol. **9** (2012)
3. Box, G.E.P., Lucas, H.L.: Design of experiments in non-linear situations. Biometrika **46**(1/2), 77–90 (1959)
4. Burrage, K., Hegland, M., Macnamara, F., Sidje, B.: A krylov-based finite state projection algorithm for solving the chemical master equation arising in the discrete modelling of biological systems. In: Proceedings of the Markov 150th Anniversary Conference, pp. 21–38. Boson Books (2006)
5. Gillespie, D.T.: Exact stochastic simulation of coupled chemical reactions. J. Phys. Chem. **81**(25), 2340–2361 (1977)
6. Henzinger, T.A., Mateescu, M., Wolf, V.: Sliding window abstraction for infinite markov chains. In: Bouajjani, A., Maler, O. (eds.) CAV 2009. LNCS, vol. 5643, pp. 337–352. Springer, Heidelberg (2009)
7. Komorowski, M., Costa, M.J., Rand, D.A., Stumpf, M.P.H.: Sensitivity, robustness, and identifiability in stochastic chemical kinetics models. Proc. Nat. Acad. Sci. **108**(21), 8645–8650 (2011)

8. Ljung, L.: System Identification: Theory for the User, 2nd edn. Prentice Hall PTR, Upper Saddle River (1998)
9. Loinger, A., Lipshtat, A., Balaban, N.Q., Biham, O.: Stochastic simulations of genetic switch systems. Phys. Rev. E **75**, 021904 (2007)
10. Mateescu, M., Wolf, V., Didier, F., Henzinger, T.A.: Fast adaptive uniformisation of the chemical master equation. IET Syst. Biol. **4**(6), 441–452 (2010)
11. Merlé, Y., Mentré, F.: Bayesian design criteria: computation, comparison, and application to a pharmacokinetic and a pharmacodynamic model. J. Pharmacokinet. Biopharm. **23**(1), 101–125 (1995)
12. Munsky, B., Khammash, M.: The finite state projection algorithm for the solution of the chemical master equation. J. Chem. Phys. **124**, 044144 (2006)
13. Pronzato, L., Walter, E.: Robust experiment design via stochastic approximation. Math. Biosci. **75**(1), 103–120 (1985)
14. Reinker, S., Altman, R.M., Timmer, J.: Parameter estimation in stochastic biochemical reactions. IEEE Proc. Syst. Biol **153**, 168–178 (2006)
15. Ruess, J., Milias-Argeitis, A., Lygeros, J.: Designing experiments to understand the variability in biochemical reaction networks. J. R. Soc. Interface **10**(88), 20130588–20130588 (2013). arXiv:1304.1455 [q-bio]
16. Sidje, R., Burrage, K., MacNamara, S.: Inexact uniformization method for computing transient distributions of Markov chains. SIAM J. Sci. Comput. **29**(6), 2562–2580 (2007)
17. van den Bos, A.: Parameter Estimation for Scientists and Engineers. Wiley-Interscience, Hoboken (2007)

# Exploring Synthetic Mass Action Models

Oded Maler[1]([⊠]), Ádám M. Halász[2], Olivier Lebeltel[1], and Ouri Maler[2]

[1] VERIMAG, CNRS, University of Grenoble-Alpes, Grenoble, France
oded.maler@imag.fr
[2] Department of Mathematics, West Virginia University, Morgantown, USA

**Abstract.** In this work we propose a model that can be used to study the dynamics of mass action systems, systems consisting of a large number of individuals whose behavior is influenced by other individuals that they encounter. Our approach is rather synthetic and abstract, viewing each individual as a probabilistic automaton that can be in one of finitely many discrete states. We demonstrate the type of investigations that can be carried out on such a model using the *Populus* toolkit. In particular, we illustrate how sensitivity to initial spatial distribution can be observed in simulation.

## 1 Introduction

*Mass action* is a fundamental notion in many situations in Chemistry, Biochemistry, Population Dynamics and Social Systems [2]. In this class of phenomena, one has a large population of individuals partitioned into several types of "species", whose dynamics is specified by a set of reaction rules. Each reaction indicates the transformation that is likely to take place when individuals of specific types come into contact. For example, a rule of the form

$$A + B \quad \rightarrow \quad A + C$$

says that when an instance of $A$ meets an instance of $B$, the latter is transformed into $C$. Denoting by $n_A$ and $n_B$ the number of instances of $A$ and $B$ existing at a certain moment, the likelihood of an $(A, B)$-encounter is proportional to $n_A \cdot n_B$. Hence the rate of change of $n_B$ will have a *negative* contribution proportional to $n_A \cdot n_B$ and that of $n_C$ will have the same magnitude of *positive* contribution. Combining for each of the species the negative contributions due to reactions in which it is transformed into something else with the positive contributions due to reactions that yield new instances of it, one can obtain a system of *polynomial*[1] differential/difference equations.

---

This work was partially supported by the French ANR project Cadmidia and the NIH Grants K25 CA131558 and R01 GM104973. Part of the work done while the second author was visiting CNRS-VERIMAG. This paper is an extended version of [14].

[1] Actually *bilinear* if one assumes the probability of triple encounters to be zero, as is often done in Chemistry.

© Springer International Publishing Switzerland 2015
O. Maler et al. (Eds.): HSB 2013 and 2014, LNBI 7699, pp. 97–110, 2015.
DOI: 10.1007/978-3-319-27656-4_6

Hybrid systems research led in the past to interactions between several branches of Computer Science and Control that have resulted in new ways to specify and analyze the behavior of complex dynamical systems [12,13]. The present paper is a preliminary step in a research program, initially inspired by [5], to pull into the CS sphere of ideas, additional domains currently dominated by the culture of Applied Mathematics and Scientific Computing, most notably the modeling and simulation of chemical reactions inside the cell [1,4,8,9].

Our approach is *top-down* and *synthetic* in the sense of defining a class of *general* mathematical models for such systems, inspired by common knowledge on the way chemical reactions work but still abstracting away from many problem-specific details due to Chemistry, Physics and even some Geometry. We believe that better conceptual and computational insights can be achieved on cleaner models, focused on what we view as essential features of the phenomenon, before adding the additional details associated with each concrete problem. We hope that investigating such models will eventually lead to novel ways to simulate and control mass action systems with potential applications, among others, in drug design and social engineering. These issues have been studied, of course, for many years in various contexts and diverse disciplines, [3,11] to mention a few, but we hope, nevertheless and despite the present apparently naive beginning, to provide a fresh look at the subject.

The rest of this paper is organized as follows. In Sect. 2 we present the basic model of the individual agent (particle) as a *probabilistic automaton* capable of being in one out of several states, and where transition labels refer to the state of the agent it encounters at a given moment. In Sect. 3 we discuss several ways to embed these individual agents in a model depicting the evolution of a large ensemble of their instances. In Sect. 4 we describe three such aggregate models, starting with a rather standard model where state variables correspond to the relative concentrations of particle types. Such models depict the dynamics of the *average* over all behaviors and they are typically realized by ordinary differential equations (ODEs) but we prefer to work in discrete time.

The second model is based on stochastic simulation under the well-stirred assumption. The third model embeds the particles in space where they perform some type of random motion and encounters may occur when the distance between two particles becomes sufficiently small. The model thus obtained is essentially a kind of a *reaction-diffusion* model for a restricted class of reactions. In Sect. 5 we briefly describe the **Populus** tool kit that we developed for exploring the dynamics of such models and illustrate its functionality. In particular, we demonstrate the effects of the initial spatial distributions of certain particle types which result in deviations from the behavior predicted by a well-stirred version of the model.

## 2    Individuals

We consider population-preserving mass action systems where new individuals are not born and existing ones neither die nor aggregate into compound entities:

they only change their internal *state*. A particle can be in one of finitely-many states and its (probabilistic) dynamics depicts what happens to it every time instant, either spontaneously or upon encountering another particle. The object specifying a particle is a probabilistic automaton:[2]

**Definition 1 (Probabilistic Automaton).** *A probabilistic automaton is a triple $\mathcal{A} = (Q, \Sigma, \delta)$ where $Q$ is a finite set of states, $\Sigma$ is a finite input alphabet and $\delta : Q \times \Sigma \times Q \to \mathbb{R}$ is a probabilistic transition function such that for every $q \in Q$ and $a \in \Sigma$,*

$$\sum_{q' \in Q} \delta(q, a, q') = 1.$$

In our model $Q = \{q_1, \ldots, q_n\}$ is the set of particle types and each instance of the automaton is always in one of those. The input alphabet is $Q \cup \{\bot\}$ intended to denote the type of *another* particle encountered by the automaton and with the special symbol $\bot$ indicating a non-encounter. An entry $\delta(q_1, q_2, q_3)$ specifies the probability that an agent of type $q_1$ converts to type $q_3$ given that it encounters an agent of type $q_2$. Likewise $\delta(q_1, \bot, q_3)$ is the probability of converting into $q_3$ spontaneously without meeting anybody. We use the notation $q_1 \xrightarrow{q_2} q_3$ for an actual invocation of the rule, that is, drawing an element of $Q$ according to probability $\delta(q_1, q_2, .)$ and obtaining $q_3$ as an outcome.

Table 1 depicts a 3-species probabilistic automaton. Looking at the diagonal of the $\bot$ matrix we can observe that the three species are rather stable in isolation. On the other hand, they may influence each other significantly upon encounter. For instance, $q_3$ transforms $q_1$ to $q_3$ with probability 0.3 while $q_1$ transforms $q_2$ to $q_3$ and $q_3$ to $q_1$ with probabilities 0.4 and 0.7, respectively.

**Table 1.** A 3-species probabilistic automaton.

| $\delta$ | $\bot$ | | | $q_1$ | | | $q_2$ | | | $q_3$ | | |
|---|---|---|---|---|---|---|---|---|---|---|---|---|
| | $q_1$ | $q_2$ | $q_3$ | $q_1$ | $q_2$ | $q_3$ | $q_1$ | $q_2$ | $q_3$ | $q_1$ | $q_2$ | $q_3$ |
| $q_1$ | 0.9 | 0.1 | 0.0 | 1.0 | 0.0 | 0.0 | 0.7 | 0.2 | 0.1 | 0.7 | 0.0 | 0.3 |
| $q_2$ | 0.1 | 0.8 | 0.1 | 0.0 | 0.6 | 0.4 | 0.0 | 1.0 | 0.0 | 0.1 | 0.9 | 0.0 |
| $q_3$ | 0.0 | 0.0 | 1.0 | 0.7 | 0.0 | 0.3 | 0.3 | 0.4 | 0.3 | 0.0 | 0.0 | 1.0 |

Our models are *synchronous* with respect to time: time evolves in fixed-size steps and at every step each particle detects whether it encounters another (and of what type) and takes the appropriate transition. The definition of when an agent meets another depends, as we shall see, on additional assumptions on the global aggregate model.

---

[2] A probabilistic automaton [15] is a Markov chain with an input alphabet where each input symbol induces a different transition matrix. It is also known as a Markov Decision Process (MDP) in some circles.

**Remark:** It is worth noting that we restrict ourselves to reaction rules which are "locally causal" in the following sense: when an $(A, B)$-encounter takes place at time $t$, the state of $A$ at time $t+1$ is does not depend on the state of $B$ at $t+1$ and vice versa: states of particles at time $t+1$ depend only on states at $t$. Compared to more general probabilistic rewrite rules that can specify the outcome of an $(A, B)$-encounter, our formalism can express rules which are products of simple rules. For instance, in a general rule like

$$A + B \rightarrow A_1 + B_1 \ (p_{11}) \mid A_1 + B_2 \ (p_{12}) \mid A_2 + B_1 \ (p_{21}) \mid A_2 + B_2 \ (p_{22})$$

the probabilities of the four outcomes should sum up to 1 while in our formulation they should satisfy the additional condition $p_{11}/p_{12} = p_{21}/p_{22}$. This restriction is not crucial for our approach but it simplifies some calculations.

## 3    Aggregation Styles

Consider now a set $S$ consisting of $m$ individuals put together, each being modeled as an automaton. The set of all possible global configurations of the system (micro-states in Physpeak) is the set $Q^S$ of all functions from $S$ to $Q$. This is an enormous state space of size $n^m$. A very useful and commonly-used abstraction is the *counting abstraction* obtained by considering two micro-states equivalent if they agree on the number of particles of each type, regardless of their particular identity. The equivalence classes of this relation form an abstracted state-space $P$ of macro-states (also known as particle number representation) each being an $n$-dimensional vector:

$$P = \{(X_1, \ldots, X_n) : \forall i \ 0 \le X_i \le m \wedge \sum_{i=1}^{n} X_i = m\}.$$

Models that track the evolution of an ensemble of particles are often viewed as dynamical systems over this abstract state-space.

For our purposes we classify models according to two features: (1) Individual vs. average dynamics; (2) Spatially-extended vs. non-spatial (well-stirred) dynamics. For the first point, let us recall the trivial but important fact that we have a non-deterministic system where being in a given micro-state, each particle tosses one or more coins, properly biased according to the states of the other particles, so as to determine its next state. To illustrate, consider a rule which transforms a particle type $A$ into $B$ with probability $p$. Starting with $m$ instances of $A$, there will be $m$ coin tosses each with probability $p$ leading to some number close to $m \cdot p$ indicating how many $A$'s convert into $B$'s. Each individual run will yield a different number (and a different sequence of subsequent numbers) but on the average (over all runs) the number of $A$'s will be reduced in the first step from $m$ to $m \cdot (1 - p)$.

Individualistic models, those used in stochastic simulation algorithms (SSA), generate such runs, one at a time. On the other hand, "deterministic" ODE models compute at every step the average number of particles for each type where this average is taken in parallel over all individual runs. For well-behaving systems, the relationship between this averaged trajectory and individual runs is of great similarity: the evolution in actual runs will appear as fluctuating around the evolution of the average. On the other hand, when we deal with more complex systems where, for example, trajectories can switch into two or more distinct and well-separated equilibria, the behavior of the average is less informative, especially when the number of molecules is small. There is a whole research thread, starting with [6], that feeds on this important distinction (see for example, [10, 16] for further discussions).

The other issue is whether and how one accounts for the distribution of particles in space. Ignoring the spatial coordinates of particles, the probability of a particular type of encounter depends only on the total number of particles of each type, which is equivalent to the well-stirred assumption: all instances of each particle type are distributed uniformly in space and hence all particles will see the same proportion of other particles in their neighborhood. This is very convenient computationally because we can work directly on the abstract state-space of particle counts. On the other hand, in spatially extended models each particle is endowed with a location which changes quasi-randomly and what it encounters along its moving neighborhood determines the interactions it is likely to participate in. Such a particle will be exposed to what happens locally along its trajectory rather than to the global number of particles. Note that embedding particles in an Euclidean-like space is just one possibility and one can think or putting them in more abstract graph-based space where distance between locations is defined by the length of shortest path, as is common in social models.

## 4    Implemented Aggregate Models

We will now describe in some detail the derivation of three models: average dynamics, individual well-stirred dynamics and spatially-extended dynamics. All our models are in discrete time which will hopefully make them more accessible to those for whom the language of integrals is not native. For the others, note that our model corresponds to a fixed time-step simulation.

### 4.1    Average Well-Stirred Dynamics

To develop the average dynamics under the well-stirred assumption we normalize the global macro-state of the system, a vector $X = (X_1, \ldots, X_n)$, into $x = (x_1, \ldots, x_n)$ with $x_i = X_i/m$ and hence $\sum x_i = 1$ (population fractions). Let $\alpha$, $0 \leq \alpha \leq 1$ be a density parameter which determines the probability of encountering another particle in one step. The evolution in this state space over time is the outcome of playing the following protocol at every time step. First, $\alpha \cdot m$ of the particles on average encounter others and hence follow a binary reaction rule while the remaining $(1 - \alpha) \cdot m$ particles do not interact and hence follow the solitary transition function. The dynamics is of the general form[3]

$$x' = x + \Delta(x),$$

where for each variable, the additive change can be written as

$$\Delta(x_k) = (1 - \alpha)\Delta_1(x_k) + \alpha\Delta_2(x_k)$$

where

$$\Delta_1(x_k) = \sum_{i=1}^{n}(x_i \cdot \delta(q_i, \perp, q_k) - x_k \cdot \delta(q_k, \perp, q_i))$$

$$\Delta_2(x_k) = \sum_{i=1}^{n}\sum_{j=1}^{n}(x_i x_j \cdot \delta(q_i, q_j, q_k) - x_k x_i \cdot \delta(q_k, q_i, q_j))$$

Here, $\Delta_1$ and $\Delta_2$ are the expected net contributions to $x_k$ by the solitary (resp. binary) reactions, each summing up the transformations of other agents into type $k$ minus the transformation of type $k$ into other types. Thus, we obtain a discrete-time bilinear dynamical system, which is linear when $\alpha = 0$.

Taking the particle automaton of Table 1 and deriving the dynamics for the sparse situation where $\alpha = 0.1$, we obtain

$$\begin{aligned}
x_1' &= x_1 - 0.09x_1 + 0.09x_2 - 0.06x_1x_2 + 0.08x_1x_3 + 0.08x_2x_3 \\
x_2' &= x_2 + 0.09x_1 - 0.18x_2 - 0.04x_1x_2 + 0.06x_2x_3 \\
x_3' &= x_3 + 0.09x_2 + 0.1x_1x_2 - 0.08x_1x_3 - 0.14x_2x_3
\end{aligned} \tag{1}$$

Starting from initial state $x = (0.4, 0.3, 0.3)$ and following the dynamics, the system converges to the state $(0.366, 0.195, 0.437)$. The same individual model, in a dense situation characterized by $\alpha = 0.9$, yields

$$\begin{aligned}
x_1' &= x_1 - 0.01x_1 + 0.01x_2 - 0.54x_1x_2 + 0.72x_1x_3 + 0.72x_2x_3 \\
x_2' &= x_2 + 0.01x_1 - 0.02x_2 - 0.36x_1x_2 + 0.54x_2x_3 \\
x_3' &= x_3 + 0.01x_2 + 0.9x_1x_2 - 0.72x_1x_3 - 1.26x_2x_3
\end{aligned} \tag{2}$$

This system, started in the same initial state $x = (0.4, 0.3, 0.3)$, converges to state $(0.939, 0.027, 0.033)$. Let us point out once more that this deterministic dynamics tracks the evolution of the average population fraction of particles over all individual runs.

## 4.2   Individual Well-Stirred Dynamics

The second model, whose average behavior is captured to some extent by the previous one, generates individual behaviors without spatial information following Algorithm 1. A micro-state of the system is represented as a set $L$ of particles,

---

[3] We export the primed variable notation from program verification where $x$ stands for $x[t]$ and $x'$ denotes $x[t + 1]$.

each denoted as $(g, q)$ where $g$ is the particle identifier and $q$ is its current state. The update round of the algorithm consists of repeatedly choosing a particle $g$ and deciding probabilistically whether it interacts with another particle. If this is the case, another particle $g'$ is randomly selected, and they both undergo their respective probabilistic binary reactions. Otherwise, $g$ is subject to a unary reaction. Reacting particles are removed from the list and the process terminates when the list becomes empty. Other variants of this procedure are mentioned in the next section.

**Algorithm 1 (Individual Well-Stirred Dynamics)**
**Input**: *A list $L$ of particles and states*
**Output** *A list $L'$ representing the next micro-state*

$L' := \emptyset$
**repeat**
   **draw** *a random particle $(g, q) \in L$; $L := L - \{(g, q)\}$*
   **draw** *binary/solitary with probability $\alpha$*
   **if** *solitary* **then**
      *apply solitary rule $q \xrightarrow{\perp} q'$*
      $L' := L' \cup \{(g, q')\}$
   **else**
      **draw** *a random particle $(g', q') \in L$; $L := L - \{(g', q')\}$*
      *apply binary rules $q \xrightarrow{q'} q''$ and $q' \xrightarrow{q} q'''$*
      $L' := L' \cup \{(g, q''), (g', q''')\}$
   **endif**
**until** $L = \emptyset$

After each update round, particle types are counted to create macro-states. The current description and implementation of the algorithm is at the level of micro-states and a more efficient stochastic simulation algorithm working directly on macro-states is discussed in Sect. 6.

## 4.3   Individual Spatial Dynamics

In our third model the particles are embedded in space, each particle represented as $(g, q, y)$ with $y$ being it spatial coordinates ranging over a bounded rectangle. The next state is computed in two phases that correspond to diffusion and reaction. First, each particle is displaced by a vector of random direction and magnitude (bounded by a constant $s$). For mathematical convenience reasons we use periodic boundary conditions that treat the rectangle as a torus: when a particle crosses the boundary of the rectangle it reappears on the other side.

Then for each particle we compute its set of neighbors $N$, those residing in a ball of a pre-specified *interaction radius $r$*, typically in the same order of magnitude as $s$. When $N$ is empty the particle undergoes a unary reaction, otherwise it interacts with a randomly chosen particle in $N$, as describe in Algorithm 2.

Other variants of the algorithm may differ by not taking the complementary transition $q' \xrightarrow{q} q'''$, and by not removing $g$ and $g'$ from $L$ (such variations apply also to the well-stirred model). We have observed empirically that these variations did not influence model behavior significantly. We should point out that the parameters we used so far rendered the situation rather dense with the typical distance between particles comparable to the interaction radius, frequently resulting in particles having multiple neighbors.

**Algorithm 2 (Individual Spatial Dynamics)**
**Input**: *A list $L$ of particles and states including planar coordinates*
**Output** *A list $L'$ representing the next micro-state*

$L' := \emptyset$
**foreach** *particle $(g, q, y) \in L$*
  **draw** *randomly $h \in [0, s]$ and $\theta \in [0, 2\pi]$*
  $y := y + (h, \theta)$
**endfor**
**repeat**
  *draw $(g, q, y) \in L$*
  $L := L - \{(g, q, y)\}$
  $N := \{(g', q', y') \in L : d(y, y') < r\}$
  **if** $N = \emptyset$ **then**

    *apply solitary rule $q \xrightarrow{\perp} q'$*
    $L' := L' \cup \{(g, q', y)\}$
  **else**
    *draw $(g', q', y') \in N$*
    $L := L - \{(g', q', y')\}$
    *apply binary rules $q \xrightarrow{q'} q''$ and $q' \xrightarrow{q} q'''$*
    $L' := L' \cup \{(g, q'', y), (g', q''', y')\}$
  **endif**
**until** $L = \emptyset$

The connection between this model, embedded in a rectangle of area $w$, and the non-spatial ones can be made via the estimation of the density factor $\alpha$. The probability of a particle $g$ not interacting with another particle $g'$ is the probability of $g'$ being outside an interaction ball, that is, $\beta = (\pi r^2)/w$, and the odds of $g$ not interacting with any of the other $m - 1$ particles is $(1 - \beta)^{m-1}$. After the reaction the number of remaining particles is either $m - 1$ or $m - 2$ and a good estimation of the average probability to interact is:

$$\alpha \approx 1 - \frac{\sum_{i=1}^{m-1}(1 - \beta)^i}{m} = 1 - \frac{1 - (1 - \beta)^m}{m\beta}.$$

## 5    The Populus Toolkit: Preliminary Experiments

We developed a prototype tool called **Populus**, written in Java and Swing, for exploring such dynamics. The input to the tool is a particle automaton

(Definition 1, Table 1) along with additional parameters such as the dimensions of the rectangle where particles live, the geometric step size $s$, the interaction radius $r$ and the initial number of each particle type, possibly restricted to some sub-rectangles. The tool simulates the three models, average well-stirred, individual well-stirred and spatially-extended, plots the evolution of particle counts over time and animates the spatial evolution. We illustrate below the type of exploration made available by the tool.

## 5.1   Average and Individual Dynamics

Figure. 1 compares the behaviors of the averaged model (deterministic) and the individual model (stochastic simulation) for two systems: a simple one where the two models exhibit similar behaviors and a more complex one where the average behavior stabilizes rapidly while individual trajectories, well-stirred and spatial, fluctuate.

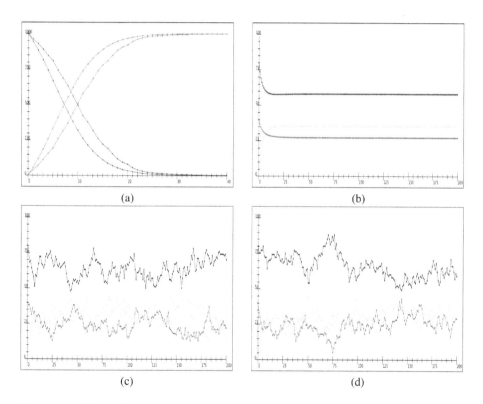

**Fig. 1.** (a) The evolution of a simple system where $A$ is eventually transformed to $B$. The smoother curves depict the average while the other curve shows an individual trajectory. (b–d) A 3-species system where the average trajectory (b) stabilizes rapidly while individual well-stirred (c) and spatial (d) trajectories fluctuate.

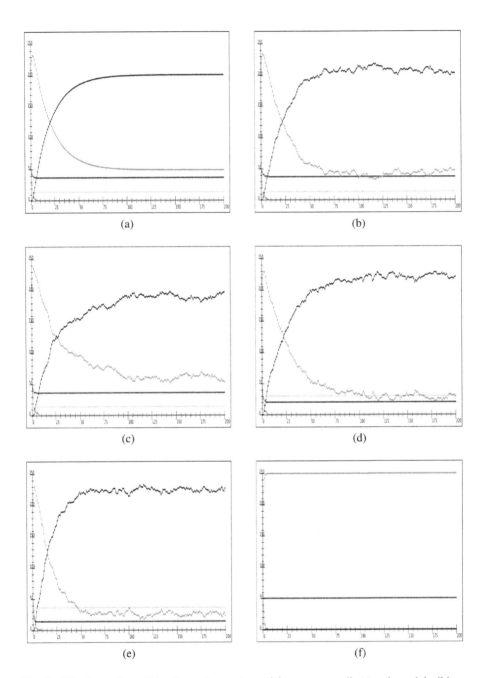

**Fig. 2.** The behavior of the 5-species system: (a) average well-stirred model; (b) an individual well-stirred run; (c) a run of the spatial model scenario 1; (d,e) two runs of scenario 2; (f) a run of scenario 3.

## 5.2   Spatial and Well-Stirred

To demonstrate the difference between spatial and non-spatial models we simulated a system with 5 species, $A$ to $E$. Particle type $A$ does not change in any transition except when it meets $B$ and converts into $C$ in certainty: $\delta(A, B, C) = 1$ and $\delta(A, \cdot, A) = 1$ otherwise. Particle type $B$ is unstable and each step it transforms into $D$ with probability 0.5: $\delta(B, \cdot, B) = \delta(B, \cdot, D) = 0.5$. Consequently it is destined to disappear after some time. Particle type $C$ is fully stable: $\delta(C, \cdot, C) = 1$. It converts $D$, which is stable under all other interactions, into $E$: $\delta(D, C, E) = 1$ and $\delta(D, \cdot, D) = 1$, otherwise. Particle type $E$ is almost stable and it transforms back into $D$ with probability 0.01: $\delta(E, \cdot, E) = 0.99$ and $\delta(E, \cdot, D) = 0.01$.

The logic behind this example is the following: there is a transient phase until $B$ disappears completely, leaving behind a number of $C$'s equal to the number of $(A, B)$ encounters that have occurred. The number of the non-converted $A$'s remains constant thereafter. Then, there are transformations of $D$ to $E$ that depend on the number of $C$'s, and interaction-independent transformations in the opposite direction. Such a system is naturally sensitive to the initial spatial distribution of $A$ and $B$ that will determine how many $C$'s will be eventually produced. We simulated the model starting from an initial state where there are 50 instances of $A$, 50 instances of $B$ and 200 instances of $D$. We used a $20 \times 20$ square over which $D$ was distributed uniformly under three scenarios that differ in the initial distribution of $A$ and $B$ in space:

- Scenario 1: $A$ and $B$ are distributed uniformly all over space;
- Scenario 2: $A$ and $B$ are concentrated initially in a unit square in the middle;
- Scenario 3: $A$ and $B$ are concentrated inside distinct unit squares far apart from each other.

The results of some runs for these scenarios are shown in Fig. 2. As a first observation, the behavior of the average model (a) and a well-stirred stochastic simulation (b) are quite similar. The number of $C$'s produced in these models with $\alpha = 0.9$ is around 13. The results of the spatial simulation of scenario 1 (c) are quite similar. In the two runs of scenario 2 (d,e), due to the proximity of $A$ and $B$ there is a burst of $(A, B)$ encounters at the beginning leading to 30 and 36 instances of $C$ and to higher levels of $E$ than under the well-stirred assumption. Finally, in scenario 3 (f) all $B$'s disappear before meeting an $A$ and hence no $C$ nor $E$ is produced.

## 6   Conclusions and Future Work

We presented a framework for studying abstract mass action dynamics based on a finite-state automaton model of the individual agent. We demonstrated how questions such as the sensitivity of some reactions to initial spatial distribution can be investigated using such models. The design philosophy underlying our framework is that all potential actions and reactions that change the state of the particle are written inside its individual model. Then the invocation of these rules by instances of the population depends on the assumptions underlying the aggregate model, its state update algorithm and the particular global state of the system.

The current framework can be improved and extended in several directions that we mention below, some being currently under investigation.

**Non Mutual Interactions and Abstract Geometries.** Our model assumed that when particles $g$ and $g'$ interact, both of them undergo reactions. In many situations such as epidemiology or social systems, the influence may work only in one way. As mentioned, this involves only minor changes in Algorithms 1 and 2 and may even simplify their probabilistic analysis. Such applications may need other notions of distance and neighborhood that reflect, for example, the separation distance between two individuals.

**Stochastic Simulation Based on Macro-States.** Algorithm 1 for well-stirred stochastic simulation works at the micro-state level, iterating over all particles and updating them one by one, resulting in $O(m)$ complexity. This fact might limit its applicability when some species exist in very large numbers. We are currently working on alternative update protocols where the determination of the set of particles that undergo binary (and unary) reactions, as well as its partitioning into pairs, are done in one preliminary step preceding the reaction step. Based on this scheme it might be easier to derive a more efficient simulation algorithm that works directly on macro-states that we sketch below (similar ideas underlie the $\tau$-leaping algorithm of [7]). The update rule for such an algorithm will have the form[4] $x' = x + \Delta(x)$ where $\Delta(x)$ is a random variable over the space of increment vectors which depends on the current macro-state. An increment vector is $\Delta = (\Delta_1, \ldots, \Delta_n)$ such that $\Delta_i \in [0, 1]$ for every $i$ and $\sum_i \Delta_i = 0$. The derivation of increment probabilities involves several steps. Assuming $\alpha$ of the particles participate in binary reactions, one needs to derive a probability over vectors $u = (u_1, \ldots, u_m)$ satisfying $\sum_i u_i = \alpha$, with $u_i$ indicating how many of those $\alpha$ particles are of type $q_i$. For each $u$ we need further to compute a probability over the different ways to partition it into pairs of particles, that is, a probability over vectors of the form $(v_{11}, \ldots, v_{nn})$ satisfying $\sum_i \sum_j v_{ij} = \alpha/2$, with $v_{ij}$ being the number of pairs $(q_i, q_j)$ that react together. This will give us probabilities over all the encounter types that, together with the rules of the automaton, can be used to derive probabilities over the increments.

**More Efficient Spatial Simulation.** The complexity of our current naive implementation of the spatial simulation algorithm is $O(m^2)$ as we need to compute the neighbors of each particle by scanning $L$. This complexity can be reduced to the necessary minimum by employing spatial data structures that

---

[4] We write the algorithm using the normalized state notation $x$ but the combinatorial calculation underlying the derivation of probabilities will be based on the particle count $X$.

can reduce the number of candidates for neighborhood that are checked. More radical performance improvements can be achieved by changing the semantics. Rather than following each particle in space we can use a more coarse-grained simulation in the spirit of the finite-element method. It consists in partitioning space into boxes and assuming each box to be well-stirred, hence represented by its (local) macro-state. The diffusion phase can be realized by particles flowing among neighboring boxes in rates proportional to their macro-state gradients. The reaction phase will consist in applying Algorithm 1 inside each box. When the size of the boxes tends to zero we obtain the current spatial algorithm and when it is the whole $W$ we have the well-stirred algorithm. Using box sizes situated between these two extremes one can define and explore models that represent the whole spectrum between the well-stirred and not stirred assumptions. Such algorithms will be more efficient than Algorithm 2 and will allow us to perform simulations with a richer range of ratios between density, velocity and interaction radius.

**Enriching the Model.** In the longer run we will consider more substantial extensions of the model of the individual. So far the movement of particles in the spatial model assumes the same speed for all particle types. This assumption seems to be relaxable without much effort. Let us also note that movement in our model is very abstract, not influenced by local densities of different particle types and it remains to be seen how such aspects can be incorporated while avoiding full-fledged kinetic simulation.

Finally, we adhere so far to population-preserving reactions where particles only change their state but do not combine together to form new entities. Given that the creation of new entities and structures is primordial for chemical and biological systems, we should look at this feature of our modeling framework as a limitation that should some day be removed. More generally, one can observe a tension between two types of models. Models of the first type are more realistic and faithful to one or more concrete physical phenomenon. They are, however full of details that may hide the forest and yield cumbersome simulation procedures. Models of the second type are cleaner and consequently are more amenable to systematic study that may even lead to some mathematical statements and general principles, at the risk of taking too much distance from any reality. We hope to make the right choices in the future and this may depend on the type and granularity of the real-life phenomena we want our models to capture.

**Acknowledgment.** We thank Eric Fanchon for many useful comments.

# References

1. Andrews, S.S., Bray, D.: Stochastic simulation of chemical reactions with spatial resolution and single molecule detail. Phys. Biol. **1**(3), 137 (2004)
2. Ball, P.: Critical Mass: How One Thing Leads to Another. Macmillan, London (2004)

3. Bortolussi, L., Hillston, J.: Checking individual agent behaviours in markov population models by fluid approximation. In: Bernardo, M., de Vink, E., Di Pierro, A., Wiklicky, H. (eds.) SFM 2013. LNCS, vol. 7938, pp. 113–149. Springer, Heidelberg (2013)
4. Burrage, K., Burrage, P.M., Leier, A., Marquez-Lago, T., Nicolau Jr., D.V.: Stochastic simulation for spatial modelling of dynamic processes in a living cell. In: Koeppl, H., Setti, G., di Bernardo, M., Densmore, D. (eds.) Design and Analysis of Biomolecular Circuits, pp. 43–62. Springer, New York (2011)
5. Cardelli, L.: Artificial biochemistry. In: Condon, A., Harel, D., Kok, J.N., Salomaa, A., Winfree, E. (eds.) Algorithmic Bioprocesses, pp. 429–462. Springer, Heidelberg (2009)
6. Gillespie, D.T.: Exact stochastic simulation of coupled chemical reactions. J. Phys. Chem. **81**(25), 2340–2361 (1977)
7. Gillespie, D.T.: Approximate accelerated stochastic simulation of chemically reacting systems. J. Chem. Phys. **115**, 1716 (2001)
8. Gillespie, D.T.: Stochastic simulation of chemical kinetics. Annu. Rev. Phys. Chem. **58**, 35–55 (2007)
9. Halász, A.M., Pryor, M.M., Wilson, B.S., Edwards, J.S.: Spatio-temporal modeling of membrane receptors (2015). under review
10. Julius, A.A., Halász, Á., Sakar, M.S., Rubin, H., Kumar, V., Pappas, G.J.: Stochastic modeling and control of biological systems: the Lactose regulation system of escherichia coli. IEEE Trans. Autom. Control **53**, 51–65 (2008)
11. Le Boudec, J.-Y., McDonald, D., Mundinger, J.: A generic mean field convergence result for systems of interacting objects. In: QEST, IEEE, pp. 3–18 (2007)
12. Maler, O.: Control from computer science. Ann. Rev. Control **26**(2), 175–187 (2002)
13. Maler, O.: On under-determined dynamical systems. In: EMSOFT, pp. 89–96. ACM (2011)
14. Maler, O., Halász, A.M., Lebeltel, O., Maler, O.: Exploring the dynamics of mass action systems. EPTCS **125**, 84–91 (2013)
15. Paz, A.: Introduction to Probabilistic Automata. Academic Press, Orlando (1971)
16. Samoilov, M.S., Arkin, A.P.: Deviant effects in molecular reaction pathways. Nat. Biotechnol. **24**(10), 1235–1240 (2006)

# Exploiting the Eigenstructure of Linear Systems to Speed up Reachability Computations

Alexandre Rocca[1,2]([✉]), Thao Dang[1], and Eric Fanchon[2]

[1] VERIMAG/CNRS, 2, Avenue de Vignate, 38610 Giere, France
alexandre.rocca@imag.fr
[2] TIMC-IMAG, UMR 5525, UJF-Grenoble 1/CNRS, 38041 Grenoble, France

**Abstract.** Reachability analysis has recently proved to be a useful technique for analysing the behaviour of under-specified biological models. In this paper, we propose a method exploiting the eigenstructure of a linear continuous system to efficiently estimate a bounded interval containing the time at which the system can reach a target set from an initial set. Then this estimation can be directly integrated in an existing algorithm for hybrid systems with linear continuous dynamics, to speed up reachability computations. Furthermore, it can also be used to improve time-efficiency of the hybridization technique that is based on a piecewise-linear approximation of non-linear continuous dynamics. The proposed method is illustrated on a number of examples including a biological model.

**Keywords:** Reachability analysis · Linear systems · Biological systems

## 1 Introduction

Linear differential systems of the form $\dot{x}(t) = Ax(t)$, where $A$ is a $n \times n$ matrix with real coefficients, constitute an important class of differential systems for which symbolic solutions are known. They have the form $x(t) = \exp(At)x_0$, where $x_0$ is an initial condition. An option is to compute numerically the matrix exponential at each time step. Another option is to write down explicitly the analytical expressions of the components $x_i(t)$ in terms of the eigenvalues and eigenvectors of $A$. In this paper we present an approach to take advantage of the eigenstructure of the matrix $A$ to speed up reachability computations of linear systems. Furthemore, it can be applied to improve the time-efficiency of the dynamic hybridization of nonlinear systems [4].

The general idea is to use the analytical expressions of $x(t)$ to estimate the time intervals over which it is certain that the linear system from a given initial set does not reach a given fixed set. Knowing in advance that no such collision is possible over these time intervals allows avoiding some computations over these intervals, for example the intersection of the reachable set and some guard set, or even accurate computations of flowpipes (sets of trajectories). The intersection computation cost growing very fast with the number of dimensions, we need a

© Springer International Publishing Switzerland 2015
O. Maler et al. (Eds.): HSB 2013 and 2014, LNBI 7699, pp. 111–127, 2015.
DOI: 10.1007/978-3-319-27656-4_7

method to avoid those computations for complex problems. If reachability analysis is greatly used for cyber-physical applications, it is less the case for biological applications because of the complexity of most of the biological systems. However, with improvements to speed it up, reachability analysis will become a powerful tool to check properties, and evaluate the robustness of biological models.

The rest of the paper is organized in two main parts. We begin the first part by presenting some preliminaries and the algorithm to estimate a set of time intervals, called Reachability Time Domain (RTD). Some experimental results are then described. Then we adapt the method of estimating RTD to speed up the dynamic hybridization of nonlinear systems. The adaptation is applied to a biological model, which shows the usefulness of the method in terms of gain in computation time. In the last section we describe related works which also exploit the eigenstructure of linear systems, and outline some directions for future work.

## 2    Reachability Time Domain Estimation

### 2.1    Preliminaries

In this section we consider a linear differential system:

$$\dot{x}(t) = Ax(t) \tag{1}$$

where $A$ is an $n \times n$ matrix with real coefficients, and $x \in \mathbb{R}^n$. We assume that the matrix $A$ is diagonalisable in $\mathbb{C}$ or, in other words, matrix $A$ has $n$ distinct eigenvalues $\Lambda = \{\lambda_1, \ldots, \lambda_n\}$, and $n$ associated eigenvectors $V = \{v_1, \ldots, v_n\}$. If some of the eigenvalues are complex, then they occur in complex conjugate pairs $(\lambda, \bar{\lambda})$. We consider that there are $r$ real eigenvalues, and $c$ pairs of complex conjugate eigenvalues, such that $\Lambda = \{\lambda_1, \ldots \lambda_r, \lambda_{r+1}, \bar{\lambda}_{r+1}, \ldots, \lambda_{r+c}, \bar{\lambda}_{r+c}\}$ (the associated real and complex eigenvectors are indexed accordingly). Obviously, $n = r + 2c$.

In this situation (distinct eigenvalues), a basic theorem of linear algebra states that the matrix $A$ can be put in block-diagonal form with blocks not bigger than $2 \times 2$. More formally, $(v_1, \ldots, v_r, Im(v_{r+1}), Re(v_{r+1}), \ldots, Im(v_{r+c}), Re(v_{r+c}))$ is a basis of $\mathbb{R}^n$ (by abuse of language we will call it the eigenbasis of $A$), the matrix

$$P = \begin{bmatrix} v_1 \ldots v_r \ Im(v_{r+1}) \ Re(v_{r+1}) \ldots Im(v_{r+c}) \ Re(v_{r+c}) \end{bmatrix} \tag{2}$$

is invertible, and $P^{-1}AP = diag[\lambda_1, \ldots, \lambda_r, B_{r+1}, \ldots, B_{r+c}]$, where $B_j$ is a $2 \times 2$ real matrix

$$B_j = \begin{bmatrix} Re(\lambda_j) & -Im(\lambda_j) \\ Im(\lambda_j) & Re(\lambda_j) \end{bmatrix}. \tag{3}$$

The notation $diag[b_j]$ stands for a block-diagonal matrix with the elements $b_j$ ($b_j$ is either a real scalar, or a real $2 \times 2$ matrix) on the diagonal.

Now consider two convex H-polytopes[1] $Init_o$ and $R_o$ in $\mathbb{R}^n$. In the following section, $Init_o$ represents the set of initial conditions and $R_o$ the target set.

---

[1] H-polytopes are polytopes defined by a set of linear constraints.

The reachability problem we address now can be formulated as the following question: does the system (1), from the initial set $Init_o$, ever reach the target set $R_o$? If the answer to this question is yes, then the first time the system enters the target set is denoted $t_{reach}$ (see Definition 2 below).

Since we want to exploit the analytical solutions of the system (1), from now on we work in the eigenbasis. This means that the two polytopes have to be transformed as follows: $Init = P^{-1}Init_o$ and $R = P^{-1}R_o$ ($Init$ is the set of initial conditions, and $R$ the target set, expressed in the eigenbasis). The Fundamental Theorem for Linear Systems [12] states that for $x_0 \in \mathbb{R}^n$ the initial value problem for the Eq. (1) and $x(0) = x_0$ has a unique solution for all $t \in \mathbb{R}$ which is given by $x(t) = e^{At}x(0)$.

For $t \geq 0$, let $E(t) = \{e^{At}x(0) \mid x(0) \in Init\}$. With this notation: $E(0) = Init$, and $E(t) = e^{At}E(0)$. From the computational point of view, the computation of $x(t)$ can be reduced to the computation of the exponential of a matrix, and to do so numerous algorithms are known [3]. In addition it is known that the image of a convex polytope by a linear operator is a convex polytope.

**Definition 1.** *We define the reach time interval $T_{overlap}$ as the set of times $t$ for which $E(t)$ intersects with $R$ (under the condition that such an intersection occurs, otherwise $T_{overlap} = \emptyset$).*

$$T_{overlap} = \{t \mid E(t) \bigcap R \neq \emptyset\} \tag{4}$$

**Definition 2.** *If $T_{overlap}$ is not empty, we define the reachability time $t_{reach} \in \mathbb{R}^+$ as the first instant $t$ for which $E(t)$ intersects $R$.*

$$t_{reach} = min\{T_{overlap}\} \tag{5}$$

**Definition 3.** *Let $T$ be a union of disjoint time intervals. $T$ is said to be a Reachability Time Domain (RTD) if $t_{reach}$ (when it exists) does not belong to the complement of $T$, then $T$ satisfies: $T_{overlap} \neq \emptyset \implies t_{reach} \in T$. Obviously, the largest RTD in all cases is $\mathbb{R}^+$, and the smallest is $\{t_{reach}\}$ when $R$ is reachable from $Init$. Note that $T_{overlap}$ is also an RTD.*

We can now restate informally our goal as follows: we want a fast algorithm to compute a useful RTD $T$. It would be for example useless to give $\mathbb{R}^+$ as an answer. On the other hand, one could design an algorithm which computes $t_{reach}$ directly by using reachability computation, and of course this is not what we intend to do here. The idea is to perform fast computations to determine an RTD $T$. Since by construction $E(t)$ cannot intersect $R$ on the complement of $T$, then it is possible to avoid the test whether $E(t)$ intersects $R$ for all time $t$ in the complement of $T$. Thus the computation of $T$ is rewarded by avoiding heavier computations.

## 2.2   Algorithm for Reachability Time Domain Estimation

We take advantage of the fact that, as mentioned above, the matrix $P^{-1}AP$ is block-diagonal in the eigenbasis, a block being just a scalar (in the case of a real eigenvalue) or a $2 \times 2$ submatrix (in the case of a pair of complex eigenvalues). This means that the system (1) can be decomposed into smaller subsystems of 1 or at most 2 variables. Remember that we assume in this work that all eigenvalues are distinct. The principle of our method is to use the analytic expressions of the solutions, expressed in the eigenbasis, and to make simple over-approximations of the convex polytopes $E(t)$ and $R$ in order to work on 1-dimensional or 2-dimensional projections.

The algorithm is divided in three parts: first, the exploitation of the real eigenvalues; second, the radial motion associated to the complex eigenvalues; third, the rotation motion associated to the complex eigenvalues. Since the differential system is decoupled when expressed in the eigenbasis, the time information extracted from the projections are independent one from the other. One could thus choose to exploit only the information associated with the real eigenvalues (assuming there is at least one). This would provide an approximation of RTD. But of course exploiting also the information associated with the complex eigenvalues provides additional constraints and generally leads to a smaller RTD.

Recall that $\lambda_i$ for $i \in \{1, \dots, r\}$ are the real eigenvalues of $A$, and that $(\lambda_i, \bar{\lambda}_i)$, $i \in \{r+1, \dots, r+c\}$ are pairs of conjugate eigenvalues.

**Part 1.** We first extract information from the real eigenvalues. The case of complex eigenvalues (presented in the next two parts) is a generalization of the basic idea presented here.

We consider each real eigenvalue $\lambda_i$, and its associated eigenvector $v_i$. The analytic solution along this axis is: $y_i(t) = e^{\lambda_i t} y_i(0)$. Now we define $proj(P, i)$ as the function that gives the projection of the polytope $P$ on the $i^{th}$ real eigenvalue subspace (subtended by $v_i$), and we call $T_i$ the time interval during which the intervals $proj(E(t), i)$ and $proj(R, i)$ overlap. The time interval $T_i$ is defined formally by:

$$T_i = \{t \mid (e^{\lambda_i t} proj(Init, i)) \bigcap proj(R, i) \neq \emptyset\} \qquad (6)$$

The bounds $t_i^{min}$ and $t_i^{max}$ of $T_i$ ($i \in \{1, \dots, r\}$) are easily computable as we will see shortly. If $R$ is reachable from $Init$ then it is clear that the time of the first encounter $t_{reach}$ belongs to all $T_i$ (because the point of contact between the two polytopes belongs to all the projections).

We define accordingly $T^{real}$ as the intersection of all the time intervals associated with real eigenvalues:

$$T^{real} = \bigcap_{0 \leq i \leq r} T_i \qquad (7)$$

From what we have just said, $T^{real}$ is an RTD. Let us call $outer(X)$ the smallest box containing the polytope $X$. Note that, from its definition, $T^{real}$ is bounded if at least one $T_i$ is bounded. Note also that even if there is a point of contact between $outer(E(t))$ and $outer(R)$ at some time $t$, we cannot conclude that $R$ is reachable from $Init$, since working on projections amounts to over-approximating the polytopes by boxes (in the subspace subtended by the real eigenvectors). In other words, if $T^{real}$ is not empty, we cannot be sure that $R$ is reachable from $Init$. But we can be sure that if $R$ is reachable, then $t_{reach}$ cannot be outside $T^{real}$. This is true independently of the existence of complex eigenvalues. In addition, if a $T_i$ is empty then we can conclude immediately that $R$ is unreachable.

Now concerning the computation of the bounds $t_i^{min}$ and $t_i^{max}$ of $T_i$, we consider a point $y_i(0)$ belonging to $proj(Init, i)$. If the configuration is such that $y_i(t)$ moves toward $R$, and the origin 0 does not lie between $y_i(0)$ and $proj(R, i)$ then it is trivial to compute the time at which $y_i(0)$ will reach $proj(R, i)$. As an example, we consider the following case: $\lambda_i < 0$ (the trajectories in this 1-dimensional subspace converge to 0), we suppose that $proj(R, i) = [z_{i,min}, z_{i,max}]$ is strictly above 0, and $y_i(0) > z_{i,max}$. Then the entry time of this point is given by: $t_i^{min} = (1/\lambda_i) \ln(z_{i,max}/y_i(0))$, and the exit time by: $t_i^{max} = (1/\lambda_i) \ln(z_{i,min}/y_i(0))$. The logarithm is negative and consequently the computed times are positive, as expected. The key property here is the monotonicity of the function $e^{\lambda_i t}$. This is just an example and a number of cases must be considered depending on: the sign of $\lambda_i$, the relative position of $proj(E(t), i)$ and $proj(R, i)$, and the position of the origin with respect to these intervals. Depending on the case, $T_i$ may be empty (meaning that $R$ is unreachable and thus the problem is solved); it may be bounded as in the above example; or it may be semi-infinite ($[t_i^{min}, -\infty]$). The lower and upper bounds of $T^{real}$ are: $t^{lb} = \max_i\{t_i^{min}\}$ and $t^{ub} = \min_i\{t_i^{max}\}$ (if at least one $t_i^{max}$ is finite).

Consider now the case where the origin 0 belongs to the box over-approximation $outer(R)$ of the target set $R$. If there is a real eigenvalue $\lambda_i$ which is negative, then the points of $(e^{\lambda_i t} proj(Init, i))$ never exit $proj(R, i)$ after entering in it, and consequently $t_i^{max}$ is infinite. We would like to obtain a *bounded* interval which is an RTD. If at least one real eigenvalue $\lambda_i$ is positive (and 0 does not belong to $outer(Init)$), then $t^{ub}$ as defined above is finite. If all the real eigenvalues are negative more work is required to get a bounded RTD. Two subcases need to be considered when all the real eigenvalues are negative. Either 0 belongs to $R$, or 0 belongs to $outer(R)$ but not to $R$ itself (we assume here that 0 does not belong to the boundary of $R$). In the first subcase we define a box containing 0 and contained in $R$, which we call $inner1(R)$. We then apply the same method as above, just replacing the outer box by the inner box $inner1(R)$: $t_i^{inner1}$ is defined as the time at which $proj(E(t), i)$ makes the first contact with $proj(inner1(R), i)$, and $t^{inner1} = \max_i\{t_i^{inner1}\}$. If $t \geq t^{inner1}$ then at least one point of the moving polytope $E(t)$ has entered the inner box $inner1(R)$. Since it is included in $R$ this point is necessarily inside $R$. This time $t^{inner1}$ thus occurs necessarily after $t_{reach}$, and can thus be taken as an upper bound for $t_{reach}$: $t^{ub} = t^{inner1}$. In the second subcase, where 0 belongs to $outer(R) \setminus R$, we define a box containing 0, contained in $outer(R)$, and disjoint from $R$. We call $inner2(R)$ a box having these properties. The time $t_i^{incl}$ is defined as the time at which $proj(E(t), i)$ is completely included in $proj(inner2(R), i)$, and globally $t^{incl} = \max_i\{t_i^{incl}\}$. If $t \geq t^{incl}$ then the moving polytope $E(t)$ is completely included in the inner box $inner2(R)$, and due to the monotonicity property, it will always remain in it. The box $inner2(R)$ being disjoint from $R$, $R$ cannot be reached after $t^{incl}$. Consequently $t_{reach}$, if it exists, is necessarily smaller than $t^{incl}$. We conclude that $t^{incl}$ can be taken as an upper bound for $t_{reach}$: $t^{ub} = t^{incl}$.

Similar reasoning can be made if 0 belongs to $Init$ (or to $outer(Init) \setminus Init$) and all the real eigenvalues are positive (case where all the $t_i^{max}$'s are infinite). The cases are too numerous to give the details here, but in the end it is only under very special conditions that the RTD resulting from the presented method is unbounded. There are basically two classes of conditions for which the above method may not provide a bounded RTD: (i) there exists only one real eigenvalue $\lambda_i$ and it is equal to 0 (the corresponding component $y_i$ is constant); (ii) there is a projection $i$ such that 0 is at an extremity of $proj(R, i)$ and $\lambda_i$ is strictly negative (or 0 is at an extremity of $proj(Init, i)$ and $\lambda_i$ is strictly positive).

The core of this part of the algorithm is straightforward: first compute $outer(R)$; if 0 does not belong to $outer(R)$, then perform the following loop for $i \in \{1, \ldots, r\}$:

- compute $t_i^{min}$ and $t_i^{max}$;
- keep the value of this $t_i^{min}$ if it is greater than the current stored value;
- keep the value of this $t_i^{max}$ if it is smaller than the current stored value.

If 0 belongs to $R$ (resp. if 0 belongs to $outer(R) \setminus R$) and if all the real eigenvalues are negative, then compute an inner box $inner1(R)$ (resp. $inner2(R)$). Then perform a similar loop in which $t^{inner1}$ (or $t^{incl}$ depending on the case) is computed instead of $t_i^{max}$, and the maximum value is retained at each step.

**Part 2.** Now we consider pairs of complex conjugate eigenvalues $(\lambda_j, \overline{\lambda}_j)$. Each such pair is associated to a $2 \times 2$ submatrix $A_j$. A trajectory defined by this matrix (and an initial condition) in the corresponding eigenplane is a spiral, or a circle if $Re(\lambda_j) = 0$, and can be decomposed into a radial and an angular component. To exploit this decomposition we use polar coordinates and we over-approximate the sets $proj(R, j)$ and $proj(I, j)$ by sectors (interval description in a polar system). In this second part we extract time information from the radial evolution of (the projection of) moving set.

The polar coordinates of a point $x$ in the eigenplane associated to $(\lambda_j, \overline{\lambda}_j)$ are noted $(\gamma, \theta)$. $E$ being a polytope in $\mathbb{R}^n$, we define the radial part of $proj(E, j)$ by:

$$\Gamma_j(E) = \{\gamma(x) \mid x \in proj(E, j)\} \tag{8}$$

The sets $\Gamma_j(Init)$ and $\Gamma_j(R)$ are intervals and we apply the same method as in Part 1. We compute for each pair $j \in \{r+1, \ldots, r+c\}$ of complex conjugate eigenvalues the time interval $T_j$ where the sector approximation of $proj(R, j)$ is reached following the radial decomposition of the motion. If there exists a pair of eigenvalues $j$, such that $T_j$ is empty then, R is unreachable. Else, we compute $T^{rad}$ the intersection of all the $T_j$ for $j \in \{r+1, \ldots, r+c\}$. Again, if $T^{rad}$ is empty then $R$ is unreachable.

$$T_j = \{ t \mid (e^{Re(\lambda_j)t} \Gamma_j(Init)) \bigcap \Gamma_j(R) \neq \emptyset \} \tag{9}$$

$$T^{rad} = \bigcap_{r+1 \leq j \leq r+c} T_j \tag{10}$$

The upper bound of the interval $T^{rad}$ can be infinite. The conditions under which this occurs are similar to those of Part 1. If the real part of all the complex eigenvalues is equal to zero, then the point trajectories lie on a product of circles (the radii depend on the initial conditions and are constant). If in addition the intersection of this set with $R$ is non empty, then the upper bound of $T^{rad}$ is infinite.

It is clear that the set $T^{real \cap rad}$ defined as the intersection of $T^{rad}$ and $T^{real}$ is an RTD. If $T^{real \cap rad}$ is empty, then $R$ is unreachable.

$$T^{real \cap rad} = T^{rad} \bigcap T^{real} \tag{11}$$

The computation of $T^{rad}$ is similar to that of $T^{real}$ in Part 1.

**Part 3.** In this last part, we extract time information from the angular motion of the reachable set. For each complex eigenvalue pair $j \in \{ r+1, \ldots, r+c \}$ we define, $\theta_j(E(t))$ the angular representation of the projection of the polytope $E(t)$ on the complex plane (a circular arc). Then we compute $T_j^{ang}$ the union of time intervals representing all the instant $t$ for which

$$\theta_j(E(t)) \bigcap \theta_j(R) \neq \emptyset.$$

Because of the periodicity of the angular motion, we describe $T_j^{ang}$ by the first interval and the period $\pi_j$.

$$T_j^{ang} = \{ t \mid \theta_j(R) \bigcap \{ e^{Bt} x_0 \mid x_0 \in \theta_j(Init) \} \neq \emptyset \} \tag{12}$$

where

$$B = \begin{bmatrix} Re(\lambda_j) & -Im(\lambda_j) \\ Im(\lambda_j) & Re(\lambda_j) \end{bmatrix}.$$

The theoretical output is the intersection of all these unions of time intervals and $T^{real \cap rad}$:

$$T^{ang} = \bigcap_{r+1 \leq j \leq r+c} T_j^{ang} \tag{13}$$

Combining all the information, the final output is:

$$T^{final} = T^{real \cap rad} \bigcap T^{ang} \tag{14}$$

In practice, the intersection to compute $T^{ang}$ is done on the fly. It is possible, mathematically, that the periods $\pi_j$ are not commensurable. In such a case, the trajectories are quasiperiodic, and $T^{ang}$ is an infinite union of intervals.

The implementation handles only floating-point numbers and consequently this case is not considered.

We can thus compute the lower common multiple of all the periods, which will be the global period $\Pi$ of the system (note that $\Pi$ can be very large). Then, even if $T^{real \cap rad}$ is not bounded (which is a very special case), the computation of $T^{ang}$ is finite in time, and $T^{final}$ can be easily represented as a finite union of time intervals, and the period $\Pi$.

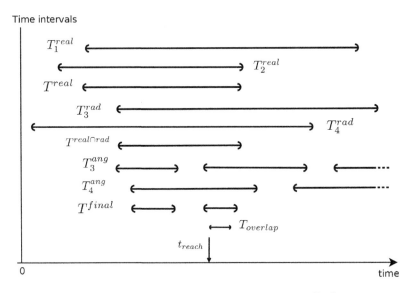

**Fig. 1.** This figure shows the different steps to construct the $T^{final}$ union of intervals for a 6-dimensional example with two real eigenvalues, and two pairs of conjugated complex eigenvalues.

## 2.3    Experimental Results

We performed two sets of experiments: the goal of the first one is to evaluate the time-efficiency of the method, and the goal of the second is to evaluate the efficiency of the method in terms of speeding up reachability computations. The experimentation was done on an Intel Pentium 4 3.60 Ghz, with 2 GBytes of memory.

The first set of experiments were carried out on randomly generated systems of dimensions 50 and 200, and the average computation times are around 4 s and 654 s respectively. The main cost of the computation comes from the computation of the box over-approximations of the initial set and of the target set.

Besides the box approximations and their projections, the computation of the lower bound of the reach time is fast (0.005 s for the systems of 200 dimensions), which shows the advantage of working on low dimensional projections.

The second set of experiments was carried out on a helicopter model with 28 variables, which is a benchmark treated by the tool SpaceEx [1]. The initial set is defined by $x_i = 0.1$ for $1 \leq i \leq 8$, and $\forall i\{1, \ldots, 28\} : x_i \in [10 - 10^{-6}, 10 + 10^{-6}]$. We searched for the time at which the system reaches a target set defined by $\forall i\{1, \ldots, 28\} : x_i \in [-2, 2]$. Our method found a reachability time at $t = 655$. This result, which is clearly smaller than the exact reach time because of the over-approximations, allowed reducing the total reachability computation time. Indeed, to compute the reachable set from the the initial up to the time point $t = 655$ SpaceEx took 397 s, while our computation of the reach time took only 0.241 s; thus we reduced the computation time by roughly $(397 - 0.241)$ s.

We can see that our method is useful in improving time-efficiency of the existing reachability algorithms, especially when the time to reach the target set from the initial set is large. In addition, to improve the accuracy of our method, the boxes may need to be subdivided, as done in [6]. Another way is to compute around the initial set the largest box that does not intersect with the target set, and then use a variant of our method for computing a lower bound on the time at which the system leaves the box. This variant is described in Sect. 3.

## 3   Application to Dynamic Hybridization

Another application of our method of reachability time domain estimation is to speed up the reachable set computation for non-linear differential systems using dynamic hybridization [4,5]. The main idea of hybridization is to construct around the current set a domain, called *hybridization domain*, within which the non-linear system is approximated by an affine system with uncertain additive input. The input here is used to account for the approximation error. When the trajectory set is inside the domain, the affine approximate system can be used to yield the analysis result for the original system with some guaranteed bounded error. To compute the reachable set of the linear approximate system inside each domain, we can use a variety of existing techniques (such as [7] and see references there in). Basically most of these techniques are based on a discretization of time into a set of consecutive small time intervals, and in each step the reachable set is approximated for the corresponding time interval. It is important to note that as soon as the trajectory set leaves the domain, this approximate system is no longer valid and a new domain and a new approximate system need to be constructed. We can see that "hybridization" here means approximating a non-linear system by a piecewise-linear system (which is a hybrid system). The hybridization technique requires therefore checking the validity of the current approximate system by testing whether the trajectory set is not entirely included in the current domain. To avoid this intersection test, we can estimate a lower bound on the first exit time, say $\tau_e$, and for any time $t < \tau_e$ the system is guaranteed to stay inside the current domain and no intersection test is needed. After the time $\tau_e$, either we stay with the current approximate system and perform intersection tests, or we construct a new hybridization domain and a new approximate system.

To estimate a lower bound on the exit time, we adapt the method for reachability time domain estimation (described in the previous section), and we then show how the time-efficiency of the reachable set computation can be enhanced by avoiding the intersection test at each step.

## 3.1  Dynamic Hybridization

First we recall the dynamic hybridization technique [4,5]. We consider the following autonomous non-linear system:

$$\dot{x}(t) = f(x(t)) \tag{15}$$

where $x \in \mathbb{X} \subseteq \mathbb{R}^n$ is the state variables and $Init \subset \mathbb{X}$ is a set of initial states.

The essential idea of the hybridization technique is as follows. It first constructs a simplicial domain $\Delta$ containing the initial set and inside $\Delta$ the dynamics $f$ is approximated by an affine system $l$. For all $x \in \Delta$:

$$l(x) = Ax + b \tag{16}$$

where $A$ is a matrix of size $n \times n$ and $b$ is a vector in $\mathbb{R}^n$. The error bound $\mu$ between the original dynamics $f$ and the approximate one, $a$, is:

$$\mu = \max_{x \in \Delta} \| f(x) - l(x) \|_\infty \tag{17}$$

This bound is used to define the input set $U_\mu \subset \mathbb{R}^n$:

$$U_\mu : \{ u \mid u \in \mathbb{R}^n \wedge \| u \|_\infty \leq \mu \} \tag{18}$$

To obtain a conservative approximate system, an input $u$ is added to the above affine system. For all $t$ such that $x(t) \in \Delta$, the non-linear system can be over-approximated by the following affine system with input:

$$\dot{x}(t) = A(x(t)) + b + u(t), u(t) \in U_\mu, x(t) \in \Delta \tag{19}$$

We denote the above system as $(\Delta, l, U)$. It is of great interest to estimate a time $\tau_e$ such that before that time: the trajectory of the approximate affine system is guaranteed to stay within the hybridization domain. To this end, we need to adapt the algorithm for reachability time domain estimation, which is explained in the next section.

## 3.2  Exit Time Prevision

From now on, we work in the transformed basis, as defined in Sect. 2.1, with the domain $\Delta$ (centered around the current set $X$) and the approximate system calculated as in (19).

To estimate a lower bound on the time at which the system intersects with the domain boundary $\partial(\Delta)$, we adapt the technique presented in the previous section. This adaptation should take into account the presence of uncertain input in the approximate dynamics. We recall that the domain $\Delta$ is a simplex.

The main idea is still to project on low dimension spaces associated with either the pairs of conjugated complex eigenvalues, or the real eigenvalues. To cope with the over-approximation due to the low dimension projection, we will use box under-approximation of the domain to stay conservative.

**Complex Eigenvalues.** For each pair of conjugate complex eigenvalues, we consider their 2-dimensional subspaces. We consider the radial evolution of the projected system to bound the exit time. To do so, we need an inner-ball approximation of the domain $\Delta$, and then we search for the time at which the current set leaves this ball, by considering the radial evolution of the system.

Let $j$ be the $j^{th}$ pair of complex conjugated eigenvalues. Let $c$ be the centroid of $Init$. We construct $B(c, \rho_b)$, the biggest ball centered at $c$ and contained in $\Delta$ and let $B_j = proj(B(c, \rho_b), j)$ be the projection of this ball on the pair $j$ of the corresponding dimensions, and $c_j = proj(c, j)$. Let $A_j$ be the matrix in this 2-dimensional system.

We are now working in a 2-dimensional subspace. We perform a translation of the coordinate subsystem so that $c_j \in \mathbb{R}^2$ becomes the origin in the new coordinate system. Let $z = y - c_j$, where $y$ is the variables of the 2-dimensional subsystem. In this new coordinate system, the dynamics of $z$ is given by:

$$\dot{z}(t) = A_j z(t) + u_c + u(t), u(t) \in U_\mu \tag{Ez}$$

where $u_c = A_j c_j$. The solution of (Ez) is:

$$z(t) = e^{A_j t} z(0) + \int_0^t e^{A_j(t-\tau)} u_c \, d\tau + \int_0^t e^{A_j(t-\tau)} u(t) \, d\tau \tag{20}$$

The exit time is the solution given by:

$$t_{exit} = min(t : ||z(t)|| \geq \rho_b)$$

To compute this time, we need a good over-approximation of $\mathcal{I} = \int_0^t e^{A_j(t-\tau)} u(t) d\tau$. To do so, we use a time discretization of step $h$ and proceed from time $t = 0$ until $||z(t)|| \geq \rho_b$. Under the uncertain input, the solution can be over-approximated by:

$$||z(t)|| \leq ||e^{A_j t} z(0) + A_j^{-1}(e^{A_j t} - I)u_c|| + ||\int_0^t e^{A_j(t-\tau)} u(t)d\tau||$$

We can prove [8] that this integral $\mathcal{I}$ for the interval $[0 ; h]$ is bounded by:

$$||\mathcal{I}|| \leq h \, ||A_j|| \, e^{h||A_j||} (2\frac{\mu}{||A_j||} + (\frac{1}{2} + h)||z(0)||). \tag{21}$$

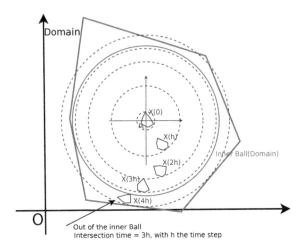

**Fig. 2.** Complex eigenvalues: we search for the intersection between the inner circle of the domain and the radial evolution of the system (in the basis centered at $u_c$). We use a stepwise computation, and in this example the intersection is found at $t = 3h$, with $h$ the time step.

**Real Eigenvalues.** Now we show how to handle real eigenvalues. The projection of the simplex on one dimension creates a large over-approximation of the domain. To keep a conservative approximation, we find a single box under-approximation $\mathcal{B} = inner(\Delta)$ of the simplex, using the algorithm in [2], and centered at the centroid of the initial set $Init$. Similarly let $\mathcal{B}_X = inner(X)$ where $X$ is the current set.

Let $proj_r$ be the operator of projecting a set on the dimensions corresponding to the real eigenvalues. Once a box under-approximation $\mathcal{B}$ of $\Delta$ is determined, we can now use the projection of $\mathcal{B}_r = proj_r(\mathcal{B})$ on each dimension associated with each real eigenvalue $\lambda_i$. Let the projection $proj(\mathcal{B}_r, i)$ be represented by two constraints $x_i \leq M$ and $x_i \geq m$ where $m, M \in \mathbb{R}$. As previously in Sect. 2.2, we can easily compute a lower bound on the exit time for the system without input, and then for the system with input, by replacing $||A_j||$ in (21) by $\lambda_i$.

### 3.3   Dynamical Hybridization with Exit Time Prevision

In the dynamic hybridization [13], the domains are dynamically constructed. Our exit time prevision method can be integrated in the hybridization algorithm to avoid polytopic inclusion tests (which in general require solving LP problems).

The main steps of the original hybridization algorithm are as follows. Given an initial $Init$. For each iteration, the algorithm performs the following steps. First, we compute an approximation domain $\Delta$ and its associated linear approximate system $(\Delta, l, U)$ as in (19). We then compute the reachable set $Rn$ from $R$ using the step-by-step algorithm with the time step $h$. We test if $newReach$ intersects with the boundary $\partial(\Delta)$ of $\Delta$. If so, we discard the set $newReach$. Otherwise, we continue with the next iteration.

Now we explain how the above-described exit time prevision method allows reducing the number of intersection tests between the set $Rn$ and the boundary $\partial(\Delta)$, by predicting a lower bound $\tau_e$ on the exit time (see Algorithm 1). If $\tau_e$ is not larger than the time step, that is $\tau_e \leq h$, we ignore this result and use the original algorithm (with intersection test at each step) until the next domain is needed. Otherwise, we can compute the reachable set for the linear approximate system $(\Delta, l, U)$ without intersection test until $\tau_e$.

---

**Algorithm 1.** Hybridisation with Exit Time Prevision.

---

```
 1: function REACH((Init, f, h))
 2:     Reach = ∅
 3:     t = 0
 4:     R = Init
 5:     repeat
 6:         (Δ, l, U) = Domain(R, f)
 7:         τe = ExitTimePrevision(R, Δ)
 8:         if (τe > h) then
 9:             /* Computing the reachable set without intersection test */
10:             for all t ≤ τe do
11:                 Rn = LinReach(R, l, U, h)
12:                 Reach = Rn ∪ Rn
13:                 R = Rn
14:                 t = t + h
15:             end for
16:         else
17:             /* Computing the reachable set with intersection test */
18:             newDomain = false
19:             repeat
20:                 Rn = LinReach(R, l, U, h)
21:                 if (Rn ∩ δ(Δ) = ∅) then
22:                     Reach = Reach ∪ Rn
23:                     R = Rn
24:                     t = t + h
25:                 else
26:                     newDomain = true
27:                 end if
28:             until newDomain
29:         end if
30:     until t ≥ tmax
31:     return Reach
32: end function
```

---

### 3.4 Experimental Results

To show how the hybridization algorithm with exit time prevision (HPA) is more time-efficiency than the original hybridization algorithm (HA) [13], we used the 7-dimensional polynomial biological model *Dictyostelium discoideum* [10], also used in [13]. This model, extracted from the molecular network, describes the aggregation stage of Dictyostellium, and its spontaneous oscillations during its development process. The equations are given bellow and the parameters value can be found in [10].

$$\frac{d[ACA]}{dt} = k1[ERK2] - k2[ACA]$$

$$\frac{d[PKA]}{d}t = k3[internal\ cAMP] - k4[PKA]$$

$$\frac{d[ERK2]}{dt} = k5[CAR1] - k6[ERK2][PKA]$$

$$\frac{d[REG\ A]}{dt} = k7 - k8[REG\ A][ERK2]$$

$$\frac{d[internal\ cAMP]}{dt} = k9[ACA] - k10[REG\ A][internal\ cAMP]$$

$$\frac{d[external\ cAMP]}{dt} = k11[ACA] - k12[external\ cAMP]$$

$$\frac{d[CAR1]}{d}t = k13[external\ cAMP] - k14[CAR1][PKA]$$

We studied the oscillating behaviours of the process. In fact, our reachability results show that for some initial states, the system can stop oscillating. This behaviour can be seen on the Fig. 3 representing the evolution of the variable $CAR1$ in function of the variable *internal cAMP*.

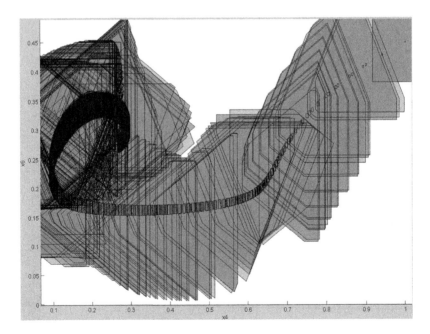

**Fig. 3.** The picture shows the projection on the plan ($CAR1$, *internal cAMP*) of the reachable set computed with [13] implementation.

Again, the experimention was done on an Intel i7 720QM quad core 1.60 Ghz, with 4 GBytes of memory. The reachable set computed by HPA are coherent with the one computed by HA, and they can be seen in Figs. 3 and 4. If we compare the total execution times, for 2000 iterations, HPA took 88 s, while HA took 128 s. The gain is 31.25%. HPA still needs intersection tests when the estimated exit time is smaller than the time step $h$ (at about 5% of the total number of iterations), but these tests took on 0.43 s, while HA needed 7.21 s for intersection tests. The total time of exit time estimation was 0.24 s.

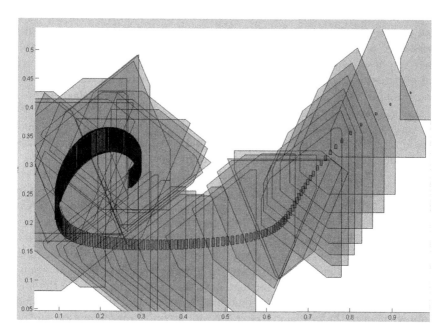

**Fig. 4.** The picture shows the projection on the plan $(CAR1, internal\ cAMP)$ of the reachable set computed with the new implementation.

Using the exit time estimation, we greatly reduced the computation time of the intersection detection. As future work, we plan to reduce the domain computation cost by using domains which are rectangular in the eigenbasis to reduce over-approximation due to the low dimension projections.

## 4   Conclusion

The essential idea of the methods presented in this paper is to extract time information from the symbolic expressions of the components $x_i(t)$ expressed in the eigenbasis of the matrix $A$ of the linear system. Our goal is to compute what we call an RTD. This allows, in a set-based simulation or a reachability analysis, to skip the parts of the time domain which corresponds to the complement of the computed RTD. Consequently it can be seen as an acceleration technique that can be integrated in reachability tools (for example SpaceEx or [13]). The authors of [9] used the symbolic expressions of the components $x_i(t)$ expressed in the eigenbasis, but in a very different way. Their goal is to define new decidable classes of linear hybrid systems. Their method, based on quantifier elimination, applies when either $A$ is nilpotent, or its eigenvalues are either all reals or all purely imaginary. Although important from the theoretical viewpoint, these classes are too restricted for practical problems. The work [11] is closer to our approach. It also uses also analytical solutions $x_i(t)$ (in the eigenbasis), and makes a piecewise linear approximation of the natural logarithm function on the real axis in order to find linear relations involving time and state variables. In that way it produces an abstraction of the solution (called time-aware relational abstraction), and then use bounded model checking to verify the linear hybrid systems. The abstraction can be refined by increasing the number of points in the piecewise linear approximation. The recent paper [6] describes a safety verification tool for linear systems also based on the idea of using symbolic expressions of the components $x_i(t)$. Their goal is to perform safety verification using a counterexample guided abstraction refinement (CEGAR) procedure. So their goal is different from ours, and consequently the way they exploit the time information contained in the $x_i(t)$ components is different, too. At the present time they use only the real eigenvalues, and plan to extend the approach to what they call quadratic eigenforms. This will allow them to extract time information from the real part of complex eigenvalues. In the present work we exploit the time information contained in real eigenvalues, and in the real and imaginary part of the complex eigenvalues. However, because of rough approximation of the initial set and target set in our methods, the application to solve the linear reachability problems suffers from some precision loss. Our future plan includes an improvement of the precision by using a more refined approximation for these sets, and by compensating the precision loss due to the projection. Combining both the dynamic hybridization technique and the linear reachability methods will give a powerful reachability tool which is also valid for non-linear systems.

**Acknowledgement.** We gratefully acknowledge the support of Agence Nationale de la Recherche (ANR) through the CADMIDIA project (grant ANR-13-CESA-0008-03).

# References

1. Spaceex: State space explorer (2010). http://spaceex.imag.fr/
2. Bemporad, A., Filippi, C., Torrisi, F.D.: Inner and outer approximations of polytopes using boxes. Computat. Geom. **27**(2), 151–178 (2004)
3. Moler, C.C., Van Loan, C.: Nineteen dubious ways to compute the exponential of a matrix, twenty-five years later. SIAM Rev. **45**(1), 3–49 (2003)
4. Dang, T., Le Guernic, C., Maler, O.: Computing reachable states for nonlinear biological models. In: Degano, P., Gorrieri, R. (eds.) CMSB 2009. LNCS, vol. 5688, pp. 126–141. Springer, Heidelberg (2009)

5. Dang, T., Maler, O., Testylier, R.: Accurate hybridization of nonlinear systems. In: Proceedings of the 13th ACM International Conference on Hybrid Systems: Computation and Control, pp. 11–20. ACM (2010)
6. Duggirala, P.S., Tiwari, A.: Safety verification for linear systems. In: 2013 Proceedings of the International Conference on Embedded Software (EMSOFT), pp. 1–10. IEEE (2013)
7. Frehse, G., Le Guernic, C., Donzé, A., Cotton, S., Ray, R., Lebeltel, O., Ripado, R., Girard, A., Dang, T., Maler, O.: SpaceEx: scalable verification of hybrid systems. In: Gopalakrishnan, G., Qadeer, S. (eds.) CAV 2011. LNCS, vol. 6806, pp. 379–395. Springer, Heidelberg (2011)
8. Guernic, C.L.: Reachability analysis of hybrid systems with linear continuous dynamics. Ph.D. thesis, Université Grenoble 1 - Joseph Fourier (2009)
9. Lafferriere, G., Pappas, G.J., Yovine, S.: A new class of decidable hybrid systems. In: Vaandrager, F.W., van Schuppen, J.H. (eds.) HSCC 1999. LNCS, vol. 1569, pp. 137–151. Springer, Heidelberg (1999)
10. Laub, M.T., Loomis, W.F.: A molecular network that produces spontaneous oscillations in excitable cells of dictyostelium. Mol. Biol. Cell 9(12), 3521–3532 (1998)
11. Mover, S., Cimatti, A., Tiwari, A., Tonetta, S.: Time-aware relational abstractions for hybrid systems. In: Proceedings of the Eleventh ACM International Conference on Embedded Software Press, p. 14. IEEE (2013)
12. Perko, L.: Linear systems. In: Differential Equations and Dynamical Systems. Springer, Heidelberg, pp. 1–63 (1991)
13. Testylier, R., Dang, T.: NLTOOLBOX: a library for reachability computation of nonlinear dynamical systems. In: Van Hung, D., Ogawa, M. (eds.) ATVA 2013. LNCS, vol. 8172, pp. 469–473. Springer, Heidelberg (2013)

# RKappa: Statistical Sampling Suite for Kappa Models

Anatoly Sorokin[1,2]([✉]), Oksana Sorokina[2], and J. Douglas Armstrong[2]

[1] Institute of Cell Biophysics RAS, Pushchino, Moscow Region, Russia
lptolik@icb.psn.ru
http://promodel.icb.psn.ru
[2] The University of Edinburgh, Edinburgh, UK

**Abstract.** We present RKappa, a framework for the development and analysis of rule-based models within a mature, statistically empowered R environment. The infrastructure allows model editing, modification, parameter sampling, simulation, statistical analysis and visualisation without leaving the R environment. We demonstrate its effectiveness through its application to Global Sensitivity Analysis, exploring it in "parallel" and "concurrent" implementations.

The pipeline was designed for high performance computing platforms and aims to facilitate analysis of the behaviour of large-scale systems with limited knowledge of exact mechanisms and respectively sparse availability of parameter values. We illustrate it here with two biological examples. The package is available on github: https://github.com/lptolik/R4Kappa.

**Keywords:** Global Sensitivity Analysis · Rule-based modeling · Model composition · Model analysis

## 1 Introduction

Dynamic modelling of biological processes is now established as a powerful tool for revealing the systems-level behaviour emerging from the interaction of molecular components. Modelling techniques based on a range mathematical grounds have been introduced over the past century, including kinetic modelling, deterministic and stochastic Petri nets, logical Boolean modelling, etc. The choice of the modelling approach generally depends upon system size, complexity, level of kinetic detail available and expected outcome. However, for a given model building task, there is no guarantee that a sufficient number of parameters are known well enough to approach biological plausibility or to ensure that the resulting simulation will be computationally tractable.

A relatively new modelling approach - rule-based modelling - is one of several developed to deal with combinatorial complexity emerging in multicomponent multistate systems [1]. It considers a complex network of interactors as a concise set of common patterns (rules) in intuitive graphical form. These have been

© Springer International Publishing Switzerland 2015
O. Maler et al. (Eds.): HSB 2013 and 2014, LNBI 7699, pp. 128–142, 2015.
DOI: 10.1007/978-3-319-27656-4_8

implemented using several different semantics (Kappa, BioNetGen, StochSim, etc.) and successfully applied to a number of well-described signalling pathways [2–4]. Rule-based modelling enables representation, simulation and analysis of the behaviour of large-scale systems where knowledge of exact mechanisms and parameters is limited. These features make it very appealing to a wide variety of biological modelling problems [5].

As an example, a routine task in bioinformatics is the construction of protein-protein interaction networks (PPINs) from a combination of proteomic and inter-actomic data. PPINs could be naturally extended by applying rule formalism to the protein-protein interactions and inferring the missing quantitative information, thus direct converting static PPI maps into a dynamic model [7]. The rule-based approach is an appropriate technique for modelling sophisticated molecular processes such as transcription and translation with highly combinatory mechanisms and limited knowledge for exact kinetic constants [6].

Rule-based generalisation of many interaction dynamics enables more effective scaling than methods that consider each interaction independently and in detail [5]. An unavoidable drawback for poor defined large-scale systems is overshooting. Local and particularly Global Sensitivity Analysis (GSA) may help resolve these issues by reducing the high-dimensional parameter space into a more tractable number of important parameters that could be measured experimentally [8]. In particular, Global Sensitivity Analysis would be of higher importance as it allows the global search of the parameter space, changing all the parameter values simultaneously to find the most sensitive parameters subset [8].

However, parameter estimation through the use of population-based global optimization techniques and consequent Local and Global Sensitivity Analysis is still a significant drain on computational resources. The combination of larger networks and the opportunity to explore parameter space rapidly demands high-performance computing platforms, such as distributed clusters, parallel supercomputers, etc.

The process of building and analysing a dynamic model generally consists of the following essential steps: model assembling, model simulation, analysis of the results and model revision. The whole process is highly iterative, therefore, an general-purpose infrastructure that supports all the steps described above would be desirable.

Indeed, for other widely used modeling techniques, such as ODE solving, a number of effective infrastructures (toolboxes) have been developed and subsequently proven their value such as COPASI, SBTOOLBOX2, SBML-SAT, SBML-PET, PottersWheel, etc. [9–13]. As an example, SBTOOLBOX2 is based around the SUNDIALS simulating engine and includes a library of Matlab scripts that support model development, model simulation, fitting of models to experimental results, parameter estimation and analysis of results, including the important options for sensitivity and identifiability analysis [11]. The SBML-SAT toolbox provides the Matlab platform for local and global sensitivity analysis [12].

For the relatively new rule-based techniques, such infrastructure is sparse in its coverage. For example, a Matlab-based library is available for BioNetGen, enabling parameter scanning, visualization and analysis of simulation results [14].

What is clearly needed is a method that facilitates the development and analysis of rule-based models within a mature statistically empowered framework. Here we present the RKappa package that embodies this need in the widely available statistical package R and demonstrates its effectiveness through its application to Global Sensitivity.

In addition to traditional GSA that we call here "parallel" for simplicity, we have introduced a computational experimental setup based upon the distinctive compositionality feature of rule-based models, which was named "concurrent" sensitivity.

We illustrate the concept with two biological models: (1) large interactomic model of postsynaptic density ("parallel" GSA) and (2) model for transcriptional initiation ("concurrent" GSA). Presented pipeline for analysis of rule-based models and the statistical evaluation of results was designed for high- performance computing platforms.

## 2    Results

### 2.1    Sensitivity Modes

We illustrate our approach to Global Sensitivity Analysis with Partial Rank Correlation Coefficient (PRCC) method [8], however it is applicable to eFAST [8], MPSA [15] and to any other algorithm that could be splitted into two distinct parts: parameter set evaluation and sensitivity coefficient calculation.

Presented setup allows running sensitivity analysis in two modes, "parallel" and "concurrent", depending on model structure and purposes (Fig. 1). In standard (we call it "parallel" here) sensitivity experiment the parameter definition part is separated from the rule, agent and observable definition part of the model with following substitution of particular parameter values for each simulation point, thus, e.g. for 10000 points sampled from parameter space we'll have to simulate 10000 models.

In biology situation when a group of similar molecules could bind another one in concurrent way is not uncommon. For example, transcription factor could bind different parts of DNA with different affinity, or phosphatase could dephosphorylate several substrates with different efficiency. Accordingly, it would be interesting to assess sensitivity of the parameters when more than one element of concurrent group is available and takes part in the interaction. We are able to perform simulation of that kind because of compositionality property of rule based models, when combination of two or more valid models is a valid model itself. Contrary to the traditional approach to the GSA ("parallel"), in "concurrent" GSA we create the single model, which consists of models obtained by substitution of parameter values from sampled points in the parameter space in the same way as in "parallel" setup, but combined together to form a supermodel. For this, the rule, agent and observable definition part of the model is additionally divided into constant and variable parts, where the constant part does not depend upon parameters varied during sensitivity analysis. The capability to generate models and simulation jobs for "concurrent" GSA is distinctive feature of our pipeline.

## 2.2 Pipeline

We selected R as an appropriate environment for developing a pipeline for several reasons. First there is a wealth of mature and readily available bioinformatics tools developed in R that are directly applicable. These include packages for data integration, analysis and visualisation. Second - R is free and widely available, which allows simple installation and immediate usage.

We focused on the rule based modelling language Kappa as it has been widely used and extended over the recent years. Several generations of Kappa simulating engines have been developed to the date, where the most recent one - KaSim3 is established as a powerful tool for modeling tasks [16]. Thus, we aimed to develop a combination of this latest generation of modelling languages, with an effective simulator all embedded in a scalable, statistical framework (R).

In our pipeline a given Kappa model undergoes the following sequence of processing steps (Fig. 1):

1. The model is loaded in R and getting prepared for further use, for example splitting into separate sections e.g. parameters, initial concentrations, rules, etc.;
2. The model is modified with respect to the future analysis, for instance, sampling of parameter space can be performed with the Sobol algorithm by specified numbers of points and parameter variation ranges; the corresponding parameter values are then applied to each model instance;
3. Simulation jobs for execution in computer cluster are created and all required data is packaged together.
4. Models are simulated with the appropriate Kappa simulator, such as KaSim [16];
5. Simulations can be run either locally or with use of parallel computation facilities depending on user demands and task size.
6. Simulation outputs are uploaded to R. Both time series of Kappa "observables" and structure of "snapshots" [16] can be analyzed (Fig. 2B). "Snapshot" structures are converted into iGraph representation for further analysis [17]. Graphs are topologically analysed with respect to specified characteristics such as size, composition, ratio of membrane/cytosol elements, etc.,
7. PRCC sensitivity coefficients are calculated by default for each characteristic and the relative parameter impact is visualised with a diagram. Other types of sensitivity metrics could be calculated with 'sensitivity' R package if required [18].

The pipeline may be executed within the R user interface or from the command line.

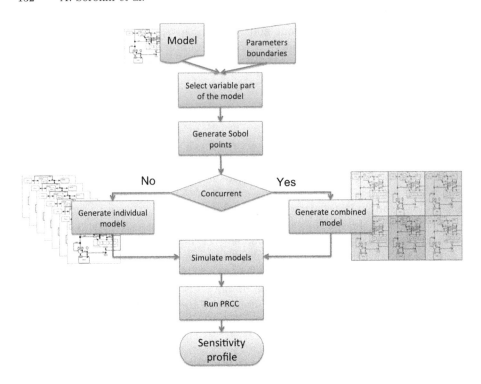

**Fig. 1.** RKappa pipeline representation. Once model and ranges of parameter values have been loaded to the pipeline the first step is separation of variable submodel from the constant part, which depends upon parameters of interest. Next step is the sampling of the parameter space. We are using Sobol low-discrepancy sequence for this, but other methods like latin square are applicable as well. The key step is preparation for the simulation: if "parallel" setup is chosen, separate model is generated and simulated for each point in the parameter space independently; for "concurrent" setup - variable submodels generated for each point in the parameter space are merged with constant part to form combined model. Generated models are simulated and analyzed as in normal PRCC algorithm.

The most time-consuming step in any population-based method, either Global Sensitivity Analysis, or parameter fitting, is evaluation of the model for particular parameter set. In case when number of parameters is high simulation of tens and even hundreds thousand of parameter sets is required. This necessitates the demand for parallelisation of such algorithms: all parameter sets could be evaluated in parallel, each on its own node. We have implemented this parallelisation strategy in our pipeline in a following form: we analyze the model and parameter space locally, generate individual model for each parameter set and prepare the jobs for simulation on high performance computing cluster; the results of the simulation could be analysed locally.

To be able to generate individual model for each parameter set we split our model into three parts :

**constant part** is the part of the model, which neither depends on nor modified by the application of the new values to the parameter set. In our toy model presented in the package vignette the only statement that belongs to that part is the snapshot definition.

**parameter part** is the definition of variables that is substituted by new parameter value assignment

**template part** while is not itself modified by the parameter assignment, still influenced by the parameter values directly, like reaction rules, or indirectly, like observables

In theory for parallel sensitivity we could keep most of the model in constant part, moving to the variable part only reaction rules, which depends upon parameters of interest, but for concurrent sensitivity, when we are going to merge many models into one metamodel, it is important to keep constant part as small as possible. Definition of the same quantity in each submodel could cause syntactic errors when metamodel is formed.

When models for each parameter set are ready we need to generate scripts to run simulations. Apparently, the pipeline is not tied rigidly to particular simulation engine, instead it requires template for the simulation script to run engine of interest on the cluster. We provide script for KaSim3 engine as default in the code, but user may define its own. The pipeline is able for validation of combination of model and engine locally so syntactic error can be fixed before submission to the calculation cluster. Below there is an example of simulation project creation:

```
proj<-prepareProject(project='model', numSets=5000, exec.path="KaSim",
  constantfiles=c('model_const.ka'), templatefiles=c("model_var.ka"),
  paramfile=c("model_param.ka"), type='parallel')
write.kproject(proj)
```

Here the new project of name "model" is created to simulate 5000 parameter sets in "parallel" GSA. Last line makes the pipeline to create folder 'model' in the working directory and write everything needed for simulation of all generated models into it. One of the generated KaSim3 simulation scripts is shown below:

```
#!/bin/bash
numEv=10
time=1000
if [ "$1" != "" ]; then
numEv= $1
echo "number of events to simulate=$numEv"
fi
if [ "$2" != "" ]; then
time= $2
```

```
echo "number of seconsd to simulate=$time"
fi
i=1
echo $i
mkdir -p "./pset43/try$i"
$KASIM_EXE  -i cABC_const.ka -i param.ka.43 -i cABC_templ.ka.43  \
  -e $time -p 100 -d "./pset43/try$i" -make-sim prom.kasim
while [ $i -lt $numEv ]
do
i=$[$i+1]
mkdir -p "./pset43/try$i"
$KASIM_EXE -e $time -p 100 -d "./pset43/try$i" -load-sim \
  ./pset43/try1/prom.kasim
done
```

To estimate sensitivity indices of stochastic models like kappa ones, Marino et al. [8] proposed to repeat simulation of each parameter set several times and analyse the average of all simulations. In the code above the number of repeated evaluations is defined by parameter "numEv".

As it was said above, generated models could be simulated either locally or remotely by running generated "job.sh" script. When simulation is completed the results could be loaded by the command:

```
abcObs<-read.observables(proj,dir='model')
```

The most common type of simulation results is observables time course. This type of data is ready for analysis straight after load. For KaSim simulator we built the additional functionality, which makes possible the analysis of the structure of interaction graph or "snapshot" obtained at the end of simulation. To perform that type of analysis kappa strings describing complexes created during simulation are converted into igraph subgraphs and combined into final snapshot graph ready for analysis.

We demonstrate our pipeline using two biological model examples as follows. The application of graph metrics for reachability analysis and analysis of parameter sensitivity in "parallel" way is demonstrated by the model of post-synaptic density, while the "concurrent" GSA is demonstrated by the transcription initiation model. Both models are available on GitHub as a part of the library documentation.

### 2.3   Kappa Model of Post-synaptic Density

Our first example is a Kappa model of the post - synaptic density (PSD) that was developed to reproduce the core structure of a large (MDa) protein complex underlying the post-synaptic membrane in mammalian neurons. The PSD is believed to mediate the major signal propagation through the synapse and its misfunctioning is thought to underlie many human diseases [19]. Proteins of the

PSD comprise a wide range of classes including scaffolds, receptors, cytoskeleton proteins and signalling enzymes [20]. They are notably enriched with specific and complimentary domains, such as PDZ and PDZ-binding C-terminal motifs, SH3, GK and some other motifs [21]. This feature was exploited for converting protein-protein interaction into a compacted list of rules based on domain-domain interaction specificity [7]. The first rule-based models of the PSD described interactions between 50 main structural proteins and reproduced sufficiently the capacity of large protein associations in the post-synaptic compartment, as well as their stability and composition variability [7].

An extended model presented here contains 89 proteins and includes signalling (phosphorylation) events in addition to protein association and dissociation processes. As previously, protein-protein interactions are formalised via 543 domain interaction and state modification rules, which require unique 124 parameters to be defined.

For each kinetic constant, values in a biologically sensible range are proposed based on the literature data. We started by sampling the parameter space from a hypercube bound by the dissociation constants (Kd) and rates of dissociation for respective protein interactions. Criteria for selection of required number of simulation points are provided in [8] and implemented within the R package. We distributed the massive computation task onto parallel computing facilities, which allowed the simultaneous exploration of 500 models. The key difficulty in simulation of such a big model was the requirement for the steady state reachability. That makes simulation is quite time consuming. In average to simulate 10 repeated evaluation of the parameter set it takes from 10 min to 2 h. Using the Eddie cluster in the Edinburgh University we were able to simulate 500 parameter sets for less then 48 h.

Upon reaching a steady state, a snapshot of the simulation was collected, parsed into R and the simulation terminated. The obtained protein complexes in the graph representation were processed and analysed with respect to their size, brutto composition, percentage of membrane elements, ratio of membrane/cytosol proteins in PSD complexes and presence and distribution of surface receptors upstream of key signalling cascades (e.g., NMDA and AMPA receptors). We also calculated the shortest paths between specific members of general signalling cascades resulting in activation of conductivity to understand how closely they are distributed in the generated models. Global sensitivity ("parallel") was calculated for each of the examined characteristics to evaluate the relative parameter impact.

For each of the 24 specified metrics we also compared "wild-type" against four simulated knock-out mutants: PSD95, SAP97, PSD93, SynGap and IRsp53. Figure 2, B shows the example sensitivity diagram for ratio AMPA/NMDA receptors, which is believed to reflect the relative strength of the synapse for the five simulated cases.

One of the main features of proteins composing the PSD is their functional redundancy, which means that in many cases the complete removal (knockout) of particular proteins will not completely disrupt key functions such as basal synaptic transmission [22]. We hypothesised that structural redundancy limits severe changes in size and structure of PSD due to compensation by the remaining proteins. This is exactly what we obtain in model simulations: all the mutated protein complexes except of the most severe case of PSD95 retain the size, ratio of membrane/cytosol proteins and AMPA/NMDA receptors similar to "wild-type".

However, in different "mutant" phenotypes different kinetic constants appear to respond for the particular characteristic performance (Fig. 2). Most importantly, length of shortest paths for members of signalling cascades varies between the different mutants and, as that length correlates with signal propagation between nodes [23], this in turn is likely to align to some degree with reported electrophysiological abnormalities observed in the *in vivo* experiments for these mutants.

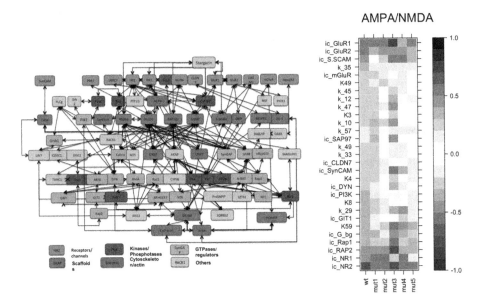

**Fig. 2.** Model of post-synaptic density. Left: structure of the model. Different functional categories of proteins are shown in respective colours. Right: Results of GSA for AMPA/NMDA ratio in the simulated PSD protein complexes are shown for 6 phenotypes (x-axes). Parameters of binding and unbinding for model components (y-axes) have different impact on the AMPA/NMDA ratio for different phenotypes, which is reflected by colour.

## 2.4   Kappa Model of Transcription Initiation

Our second proof-of-concept is a model for bacterial transcription initiation (Fig. 4). The Kappa model describes the process of promoter localisation by *E. coli RNA* polymerase and all five steps of transcription initiation. It was built upon known continuous models and correlates well with *in vitro* experimental data ([24,25]). It is generally accepted, that the number of available RNA polymerase molecules is smaller (2000-4000 per cell) than number of available

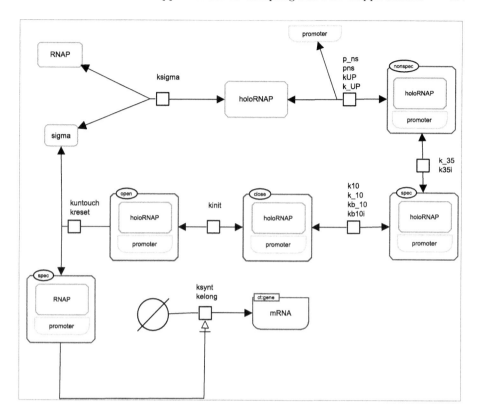

**Fig. 3.** Simplified representation of main interactions in the transcription initiation model. Kinetic constants are shown next to reaction nodes.

promoters (4500-5500), while the number of sigma-subunits that is required by RNA polymerase to bind to the promoter and initiate transcription is even less (about 800 per cell [26,27]). RNA polymerase, both by itself and in the complex with a sigma-subunit called holoenzyme, is able to bind any site of DNA in a weak nonspecific way, even though it is not able to initiate transcription from it. Therefore, *in vivo*, promoters have to compete for the active RNA polymerase molecules in the cell. That environment is in stark contrast to the situation generally applied *in vitro* experiments to measure various kinetic parameters of RNA polymerase-promoter interaction process: practically all the experimental data used to create current models of transcription initiation are obtained in conditions of the at least equimolar concentration of promoter and RNA polymerase holoenzyme. In some cases concentration of the protein is even three- to ten-fold higher than concentration of the promoter DNA. These environmental differences could conceivably result in significant underestimation of the role of processes such as promoter localisation, initial non-specific binding and the role of non-promoter DNA in transcription initiation. Our kappa model of transcription initiation was developed to allow us to explore the influence of different steps

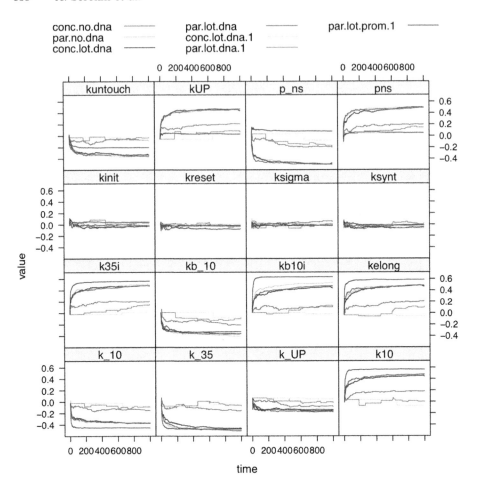

**Fig. 4.** Time dependent sensitivity of transcription initiation model at parallel and concurrent conditions. "Concurrent" setups: 104 non-specific DNA, 1 promoter, 2000 RNAP (green); 0 no-specific DNA, 1 promoter, 2000 RNAP (blue); 104 non-specific DNA, 1 promoter, 2 RNAP (orange). "Parallel" setups: 104 non-specific DNA, 1 promoter, 2000 RNAP (red); 0 no-specific DNA, 1 promoter, 2000 RNAP (magenta); 104 non-specific DNA, 1 promoter, 2 RNAP (yellow); 104 non-specific DNA, 500 promoters, 2 RNAP (pink). It could be seen that PRCC in "parallel" and "concurrent" setups are quite different, for example "parallel" setup without DNA is close to "concurrent" setup with a lot of DNA.

of transcription initiation in competition between various promoters for limited number of holoenzymes in the cell. To prove that the structure of the model is correct we took parameter values from [28] and show that the simulation results were close to the experimental data.

Here the "concurrent" mode is the best choice, as the constant part, which in our case describes RNA-polymerase itself, does not depend upon parameter values varied during sensitivity analysis. The variable part describes interaction between RNA-polymerase and promoters. The supermodel obtained can reflect more accurately the *in vivo* environment because of the large number of individual promoters characterized by various parameter sets, which allow us to model competition for interaction with limited number of enzymes.

To compare the performance of the "concurrent" and "parallel" GSA we have run our transcription initiation model with seven different combinations of RNAP, promoter and non-specific DNA concentrations. Results of time-dependent sensitivity coefficients for all computational experiments are shown on the (Fig. 3). We can see that sensitivity profiles of "parallel" GSA does not change upon change in amount of non-promoter DNA, while in "concurrent" GSA most parameters are less sensitive without non-promoter DNA.

## 3   Discussion

Over the past 10 years, rule-based modelling has established a reputation as a useful tool for molecular simulations. Amongst its benefits, its inherent ability to scale to cope with the high combinatorial complexity of biological systems is arguably the most important.

One example of such system is protein interactomics. Topological analysis of PPI maps is a well-known strategy for learning the basic principles of protein network organization; lots of studies performed up to date to identify the functionally meaningful clusters/motifs in the protein network [29–31].

The rule-based semantics endows a qualitative PPI map with the missing information essential for quantitative modelling; explicitly describing protein binding sites (including possible modifications), concentrations and affinities. Each rule defines what is essential for the particular interaction and omits the irrelevant information, thus, the plethora of concurrent modifying and binding events can be wrapped into a relatively compact executable dynamic model.

Similar reasoning can be applied to other levels of molecular processes such as transcriptional regulation. Rules describing the essential steps in the interactions between RNA-polymerases and promoters try to take into account as much experimental information as possible whereby an individual plasmid promoter or even all promoters in the cell can potentially be modelled.

The output of any mathematical model is inevitably subjected to uncertainty as the model input is a priori based on several sources of uncertainty, such as absence of information about exact parameter values, erratic measurements and simply poor or partial understanding of the process under investigation. For models built upon the sizable protein-protein interaction networks the number of undefined parameters becomes enormously large and most of them could not be identified experimentally.

Likewise, it is unlikely possible to define experimentally all the rate constants for the binding of individual promotors by RNA-polymerase. However, it is essential to understand relative role of transcription initiation steps in the control of gene expression and relate parameters important for *in vivo* performance to parameters measurable in the *in vitro* experiments.

Sensitivity analysis allows us to rank model parameters with respect to their impact within the huge parameter space. GSA in particular. That decrease in task dimensionality makes search for optimal solution much easier.

This approach is based on the assumption that the individual parameters having the biggest impact on the model are also the most important ones to optimise. Modellers will obviously struggle to find optimal solutions if they are distributed over a wide range of parameters each of which has minimal individual contribution. However, one of the key purposes of this type of modelling is to identify key nodes in molecular networks that can be measured or manipulated in biological systems where the inherent noise would overwhelm such cases.

Kappa evidently lacks a single pipeline for editing the model, configuring initial conditions, iterative modification, simulation, with analysis the results and their visualization. Existing tools do allow running simulations locally, e.g. with KaSim and provide some primary knowledge for model behaviour and structure. These approaches work perfectly well for compact models with well- defined parameters. If one needs to run the multiple versions of large-scale models or compare thousands runs for the given model to explore parameters then existing solutions struggle to cope. Such studies require much more extensive computation and parallel computing becomes a more viable option.

The pipeline we describe here facilitates automatic generation of updated versions of the rule-based models with modified kinetic rates and initial concentrations; it prepares the models for parallel/clustering facilities, simulates them with Kappa simulators (JSim or KaSim), runs GSA and then provides process simulation output with respect to user requirements and finally provides a convenient visualization of the results in a form of graphs. All this can be achieved without leaving R environment, or alternatively from command line.

The current pipeline does not attempt to duplicate the infrastructure designed for building the Kappa models, debugging them and performing initial analysis of their structure as such capabilities are well covered by solutions such as KaSim, simplx/complx, RuleStudio to name a few. Rather we concentrate on managing the simulations and processing of simulation results. The successful implementation of a comprehensive pipeline would naturally entail the design of the infrastructure for the automatic generation the Kappa models from PPi map, Boolean genetic regulatory networks or causality networks inferred from various-omics experiments. However, at the first step using of valid, manually curated models is essential for understanding of the applicability and capacity of the approach.

For both models used as examples the huge parameter space with mostly unknown exact kinetic values makes the task of getting plausible biological insights quite difficult. Global Sensitivity Analysis implemented in two ways allows significant reduction of parameters to the tractable number of most important ones, which in combination with experiments allows model to make realistic predictions.

**Acknowledgments.** AS was partially supported by RFBR, research project No. 14-44-03679 r_centr_a, and European Research Council (ERC) under grants 320823 RULE. The research leading to these results has received funding from the European Union Seventh Framework Programme (FP7/2007-2013) under grant agreement nos. 241498 (EUROSPIN project), 242167 (SynSys-project) and 604102 (Human Brain Project). This work has made use of the resources provided by the Edinburgh Compute and Data Facility (ECDF) (http://www.ecdf.ed.ac.uk/). The ECDF is partially supported by the eDIKT initiative (http://www.edikt.org.uk).

# References

1. Chylek, L.A., Harris, L.A., Tung, C.-S., Faeder, J.R., Lopez, C.F., Hlavacek, W.S.: Rule-based modeling: a computational approach for studying biomolecular site dynamics in cell signaling systems. Wiley Interdisc. Rev. Syst. Biol. Med. **6**, 13–36 (2014)

2. Danos, V., Feret, J., Fontana, W., Harmer, R., Krivine, J.: Rule-based modelling and model perturbation. In: Priami, C., Back, R.-J., Petre, I. (eds.) Transactions on Computational Systems Biology XI. LNCS, vol. 5750, pp. 116–137. Springer, Heidelberg (2009)

3. Faeder, J.R., Blinov, M.L., Hlavacek, W.S.: Rule-based modeling of biochemical systems with BioNetGen. Syst. Biol. Methods Mol. Biol. **500**, 113–167 (2009)

4. Novere, N.L., Shimizu, T.S.: STOCHSIM: modelling of stochastic biomolecular processes. Bioinformatics **17**, 575–576 (2001)

5. Danos, V., Feret, J., Fontana, W., Krivine, J.: Scalable simulation of cellular signaling networks. In: Shao, Z. (ed.) APLAS 2007. LNCS, vol. 4807, pp. 139–157. Springer, Heidelberg (2007)

6. Sorokin, A., Temlyakova, E.: Rule-based model of bacterial transcription initiation. FEBS J. **280**, 569 (2013)

7. Sorokina, O., Sorokin, A., Armstrong, J.D.: Towards a quantitative model of the post-synaptic proteome. Mol. BioSyst. **7**, 2813–2823 (2011)

8. Marino, S., Hogue, I.B., Ray, C.J., Kirschner, D.E.: A methodology for performing global uncertainty and sensitivity analysis in systems biology. J. Theor. Biol. **254**, 178–196 (2008)

9. Maiwald, T., Timmer, J.: Dynamical modeling and multi-experiment fitting with PottersWheel. Bioinformatics **24**, 2037–2043 (2008)

10. Mendes, P., Hoops, S., Sahle, S., Gauges, R., Dada, J., Kummer, U.: Computational modeling of biochemical networks using COPASI. Syst. Biol. Methods Mol. Biol. **500**, 17–59 (2009)

11. Schmidt, H., Jirstrand, M.: Systems biology toolbox for MATLAB: a computational platform for research in systems biology. Bioinformatics **22**, 514–515 (2006)

12. Zi, Z., Zheng, Y., Rundell, A.E., Klipp, E.: SBML-SAT: a systems biology markup language (SBML) based sensitivity analysis tool. BMC Bioinf. **9**, 342 (2008)

13. Zi, Z.: SBML-PET-MPI: a parallel parameter estimation tool for systems biology markup language based models. Bioinformatics **27**, 1028–1029 (2011)

14. Sneddon, M.W., Faeder, J.R., Emonet, T.: Efficient modeling, simulation and coarse-graining of biological complexity with NFsim. Nature Methods **8**, 177–183 (2010)

15. Cho, K.-H., Shin, S.-Y., Kolch, W., Wolkenhauer, O.: Experimental design in systems biology, based on parameter sensitivity analysis using a monte carlo method: a case study for the TNF?-mediated NF-? B Sign. Transduct. Pathway Simul. **79**, 726–739 (2003)
16. The Kappa Language. http://www.kappalanguage.org/
17. igraph: The network analysis package. http://igraph.org/
18. Pujol, G., Iooss, B.: Sensitivity: Sensitivity Analysis in R (2008)
19. Baron, M.K., Boeckers, T.M., Vaida, B., Faham, S., Gingery, M., Sawaya, M.R., Salyer, D., Gundelfinger, E.D., Bowie, J.U.: An architectural framework that may lie at the core of the postsynaptic density. Science **311**, 531–535 (2006)
20. Cheng, D., Hoogenraad, C.C., Rush, J., Ramm, E., Schlager, M.A., Duong, D.M., Xu, P., Wijayawardana, S.R., Hanfelt, J., Nakagawa, T., Sheng, M., Peng, J.: Relative and absolute quantification of postsynaptic density proteome isolated from rat forebrain and cerebellum. Mol. Cell. Proteomics **5**, 1158–1170 (2006)
21. Nourry, C., Grant, S.G.N., Borg, J.-P.: PDZ Domain Proteins: Plug and Play! Sci. STKE 179, re7 (2003)
22. Carlisle, H.J., Fink, A.E., Grant, S.G.N., O'Dell, T.J.: Opposing effects of PSD-93 and PSD-95 on long-term potentiation and spike timing-dependent plasticity. J. Physiol. **586**, 5885–5900 (2008)
23. Borgatti, S.P.: Centrality and network flow. Soc. Netw. **27**, 55–71 (2005)
24. Saecker Jr, R.M., M.T.R., deHaseth, P.L.,: Mechanism of bacterial transcription initiation: RNA polymerase - promoter binding, isomerization to initiation-competent open complexes, and initiation of RNA synthesis. J. MolecularBiology **412**, 754–771 (2011)
25. Liang, S.-T., Bipatnath, M., Xu, Y.-C., Chen, S.-L., Dennis, P., Ehrenberg, M., Bremer, H.: Activities of constitutive promoters in Escherichia coli. J. Mol. Biol. **2921**, 19–37 (1999)
26. Ishihama, A.: Functional modulation of escherichia coli rna polymerase. Microbiology **54**, 499–518 (2000)
27. Ishihama, Y., Schmidt, T., Rappsilber, J., Mann, M., Hartl, F.U., Kerner, M.J., Frishman, D.: Protein abundance profiling of the Escherichia coli cytosol. BMC Genomics **9**, 102 (2008)
28. Sclavi, B., Zaychikov, E., Rogozina, A., Walther, F., Buckle, M., Heumann, H.: Real-time characterization of intermediates in the pathway to open complex formation by Escherichia coli RNA polymerase at the T7A1 promoter. PNAS **102**, 4706–4711 (2005)
29. Newman, M.E.J., Girvan, M.: Finding and evaluating community structure in networks. Phys. Rev. **69**, 026113 (2004)
30. Wang, J., Li, M., Deng, Y., Pan, Y.: Recent advances in clustering methods for protein interaction networks. BMC Genomics **11**, S10 (2010)
31. Pocklington, A.J., Cumiskey, M., Armstrong, J.D., Grant, S.G.N.: The proteomes of neurotransmitter receptor complexes form modular networks with distributed functionality underlying plasticity and behaviour. Mol Syst Biol. 2, 2006.0023 (2006)

# Integration of Rule-Based Models and Compartmental Models of Neurons

David C. Sterratt[✉], Oksana Sorokina, and J. Douglas Armstrong

School of Informatics, University of Edinburgh, 10 Crichton Street,
Edinburgh EH8 9AB, Scotland, UK
{david.c.sterratt,douglas.armstrong}@ed.ac.uk, osorokin@inf.ed.ac.uk

**Abstract.** Synaptic plasticity depends on the interaction between electrical activity in neurons and the synaptic proteome, the collection of over 1000 proteins in the post-synaptic density (PSD) of synapses. To construct models of synaptic plasticity with realistic numbers of proteins, we aim to combine rule-based models of molecular interactions in the synaptic proteome with compartmental models of the electrical activity of neurons. Rule-based models allow interactions between the combinatorially large number of protein complexes in the postsynaptic proteome to be expressed straightforwardly. Simulations of rule-based models are stochastic and thus can deal with the small copy numbers of proteins and complexes in the PSD. Compartmental models of neurons are expressed as systems of coupled ordinary differential equations and solved deterministically. We present an algorithm which incorporates stochastic rule-based models into deterministic compartmental models and demonstrate an implementation ("KappaNEURON") of this hybrid system using the SpatialKappa and NEURON simulators.

**Keywords:** Hybrid stochastic-deterministic simulations · Hybrid spatial-nonspatial simulations · Multiscale simulation · Rule-based models · Compartmental models · Computational neuroscience

## 1 Introduction

The experimental phenomena of long term potentiation (LTP) and long term depression (LTD) show that synapses can transduce patterns of electrical activity on a timescale of milliseconds in the neurons they connect into long-lasting changes in the expression levels of neurotransmitter receptor proteins. This synaptic plasticity plays a crucial role in the development of a functional nervous

The research leading to these results has received funding from the European Union Seventh Framework Programme (FP7/2007-2013) under grant agreement nos. 241498 (EUROSPIN project), 242167 (SynSys-project) and 604102 (Human Brain Project). We thank Anatoly Sorokin for his help with SpatialKappa and comments on an earlier version of the manuscript, and Vincent Danos for thought-provoking discussions.

© Springer International Publishing Switzerland 2015
O. Maler et al. (Eds.): HSB 2013 and 2014, LNBI 7699, pp. 143–158, 2015.
DOI: 10.1007/978-3-319-27656-4_9

system and in encoding semantic memories (e.g. motor patterns) and episodic memories (experiences), converting stimuli lasting for seconds into memories that last a lifetime [16].

There are a number of computational models of how synaptic plasticity arises from patterns of pre- and postsynaptic electrical activity, the dynamics of $\alpha$-amino-3-hydroxy-5-methyl-4-isoxazolepropionic acid receptors (AMPARs) and $N$-methyl-D-aspartic acid receptors (NMDARs), calcium influx through these receptors and intracellular signalling in the postsynaptic density (PSD), a dense, protein-rich structure attached to neurotransmitter receptors [1,14,22,28]. The level of detail of the molecular component of these models ranges from deterministic simulations in one compartment [1] through stochastic models with coarse granularity [27] and, at the most detailed, particle-based simulations in which the Brownian motion of individual molecules is modelled [26,28].

The model with the greatest number of molecular species has 75 variables representing the concentrations of signalling molecules, complexes of signalling molecules and phosphorylation states [1]. This constitutes a small subset of the 1000 proteins identified in the mouse postsynaptic proteome, the collection of proteins in the PSD [6]. Even the subset of the postsynaptic proteome containing proteins associated with membrane-bound neurotransmitter receptors contains over 100 members [20]. As these proteins are particularly associated with synaptic plasticity, it would be desirable to increase the number of proteins and complexes it is possible to describe in simulations.

As each protein has a number of binding sites, a combinatorially large number of complexes can arise, meaning that a correspondingly large number of reactions are needed to describe the dynamics of all possible complexes. By specifying rules whose elements are fragments of complexes, rule-based languages and simulators [5], such as Kappa [7] or BioNetGen [9], obviate the need to specify reactions for all possible complexes. These rules are simulated using a method similar to Gillespie's stochastic simulation method for reactions [7]. Kappa has been used to predict the sizes of clusters of proteins in the postsynaptic proteome [23].

Compartmental models of electrical activity in neurons split the neuronal morphology into a number (ranging from 1 to around 1000) of compartments, and specify the dynamics of the membrane potential in each compartment in terms of coupled ordinary differential equations (ODEs) [12,25]. Quantities beyond the membrane potential can also be modelled, e.g. intracellular calcium concentration and concentrations of a few other molecules such as buffers and pumps. Various packages can generate and solve the equations underlying compartmental models from various model description languages, for example NEURON [3], MOOSE [21] and PSICS [2].

We present an algorithm which integrates rule-based models and compartmental models of neurons. To be sure of understanding a simple, yet interesting, case, we limit ourselves to considering isolated postsynaptic proteomes in a neuron of arbitrary morphology. Although of interest, we do not consider diffusion of molecules within the neuron. We implement the algorithm by incorporating the SpatialKappa rule-based simulator [24] into the NEURON simulator [3].

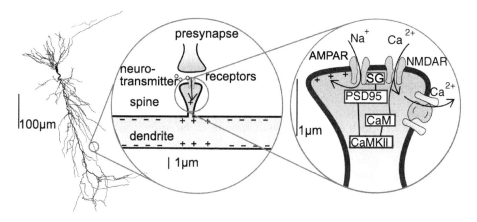

**Fig. 1.** The system to be simulated depicted on the scale of a whole neuron (left), a spine (middle) and the postsynapse (right). Scale bars are approximate. In the postsynapse, the rectangular boxes represent intracellular molecules: SG – stargazin; CaM – calmodulin; CaMKII – $Ca^2+$/calmodulin-dependent protein kinase II.

We validate the combined simulator ("KappaNEURON") against stand-alone NEURON, and demonstrate how the system can be used to simulate complex models.

## 2    Simulation Method

### 2.1    The System to be Simulated

An example of the type of system to be simulated is shown in Fig. 1. There is a hippocampal CA1 pyramidal neuron (left) upon which are located a number of synapses. Excitatory synapses are generally located on synaptic spines, small protuberances from the neuron whose narrow necks limit, to an extent, diffusion of ions and molecules between the spine head and the rest of the neuron (middle). The synapse contains a postsynaptic proteome of arbitrary complexity (right). Firing in the presynaptic neuron causes release of neurotransmitter from the presynaptic bouton, which, after diffusing across the synaptic cleft, binds to AMPARs and NMDARs. The AMPARs open and close on a submillisecond timescale, allowing sodium ions to flow into the cell. These ions charge the membrane locally and flow to other parts of the neuron, where they also charge the membrane (middle, Fig. 1). The NMDARs open and close with slower dynamics, and allow calcium ions to flow into the spine. Inside the spine, the calcium ions bind to various proteins such as calmodulin and the resulting calcium-calmodulin complex may then bind to $Ca^2+$/calmodulin-dependent protein kinase II (CaMKII), initiating signalling known to be critical for the induction of LTP and LTD.

We will first give the general set of equations that constitute a deterministic description of this type of system, then apply the general equations to a specific example, and next describe how the equations are solved. Finally we will show how we incorporate rule-based models and simulate the hybrid system.

## 2.2   Deterministic Description of Electrical and Chemical Activity in a Neuron

To describe electrical activity in the neuron, the neuron is split into compartments, each of which should be small enough to be approximately isopotential. Apart from the root compartment, which is located in the soma, each compartment has a parent, and a compartment may have one or more children. The equation for the membrane potential $V_i$ in compartment $i$ derives from Kirchoff's current law:

$$C_i \frac{dV_i}{dt} = \sum_{j \in \mathcal{N}i} \frac{d_{ij}(V_j - V_i)}{4R_a l_{ij}^2} - \sum_S \left( I_{S,i}^{chan} + I_{S,i}^{pump} \right) - I_{ns,i}^{chan} \; . \tag{1}$$

The left hand side is the current per unit membrane area charging or discharging the membrane; $C_i$ is the specific membrane capacitance in compartment $i$. The first term on the right hand side describes current flow into compartment $i$ from its neighbours $j \in \mathcal{N}_i$; $R_a$ is the intracellular resistivity, $l_{ij}$ is the path length between the midpoints of $i$ and each of its neighbours $j$, and $d_{ij}$ is the mean diameter of the path. The second term on the right hand side is the total transmembrane current per unit area (referred to as current density) in compartment $i$ carried by various species of ion $S$ via ion channels ($I_{S,i}^{chan}$) and membrane pumps ($I_{S,i}^{pump}$), which act to maintain concentration differences between the intracellular and extracellular space. To represent "non-specific" ion currents whose concentration is not accounted for, there is final term $I_{ns,i}^{chan}$. Here the minus sign reflects the conventions that inward current is negative and the extracellular space is regarded as electrical ground.

The current density carried by species $S$ through types $k$ of ion channel is:

$$I_{S,i}^{chan} = \sum_k g_{ik}(O_{ik}, t) f_{S,k}(V_i, [S]_i, [S]_o) \; , \tag{2}$$

where $g_{ik}$ is the conductance of ion channel type $k$ in compartment $i$, which may be a function of time or a state variable $O_{ik}$, and $f_{S,k}$ and is a function describing the $I$–$V$ characteristic of current flow of ions of type $S$ through channel $k$, which may depend on $[S]_i$, the intracellular concentration of $S$ in compartment $i$, and $[S]_o$, the extracellular concentration of $S$, which is assumed to be constant. A normalised Goldmann-Hodgkin-Katz (GHK) current equation [25] can be used for $f_{S,k}$. For channels through which calcium flows, the typically large ratio between intracellular and extracellular calcium concentrations means this function depends quite strongly on the intracellular calcium concentration $[Ca^{2+}]$, but in channels not permeable to calcium it is usual to use a linear approximation $V_i - E_k$, where $E_k$ is the reversal potential for that channel. By removing the dependence on intracellular concentrations $[S]_i$, this approximation allows currents carried by ions other than calcium to lumped together in a nonspecific ion category.

The state variable $O_{ik}$ is the number of type $k$ ion channels in compartment $i$ which are in an open conformation. It is modelled as the occupancy of a state of a Markov process with membrane potential-dependent transition rates. For small number of channels the Markov process is simulated stochastically, but with large numbers of channels the system is practically deterministic and the master equation of the Markov process is simulated using ODEs.

The dynamics of the intracellular concentrations of ions can be modelled using further ODEs. The rate of change of $[S]_i$ depends on $I_{S,i}^{\text{chan}}$, the channel transmembrane current density carried by $S$, and consumption and release by intracellular reactions:

$$\frac{\mathrm{d}[S]_i}{\mathrm{d}t} = -\frac{a_i}{z_S F v_i} I_{S,i}^{\text{chan}} + \sum_r J_{S,r,i} \ , \tag{3}$$

where $a_i$ is the surface area of the compartment, $v_i$ is the volume of the compartment, $z_S$ is the valency of ion $S$, and $F$ is Faraday's constant. The term $\sum_r J_{S,r,i}$ describes the net flux of $S$ due to intracellular reactions $r$. It arises from treating the intracellular reactions in compartment $i$ as a set of kinetic schemes:

$$r : S + T \xrightleftharpoons[k_{-r}]{k_r} S \cdot T \ . \tag{4}$$

The flux of $S$ arising from this reaction would be:

$$J_{S,r,i} = -k_r [S]_i [T]_i + k_{-r} [S \cdot T]_i \ . \tag{5}$$

The pump current $I_{S,i}^{\text{pump}}$ may be defined in terms of the flux of a reaction $r$, for example:

$$I_{S,i}^{\text{pump}} = \frac{z_S F v_i}{a_i} J_{S,r,i} \ , \tag{6}$$

where the prefactor converts from flux to current. Thus the whole electrical and molecular system is defined by a system of ODEs.

## 2.3   Example Deterministic Description

To help understand the formalism above, we provide an example of a simple one-compartment system; this will also be used as the validation example in Sect. 4.1. There is a single compartment whose membrane contains passive (leak) channels, calcium channels, and a transmembrane calcium pump, described by the kinetic scheme:

$$\text{Ca binding:} \quad \text{P} + \text{Ca} \xrightarrow{k_1} \text{P} \cdot \text{Ca}$$
$$\text{Ca release:} \quad \text{P} \cdot \text{Ca} \xrightarrow{k_2} \text{P} \ , \tag{7}$$

where Ca represents intracellular calcium, P represents a pump molecule in the membrane, $\text{P} \cdot \text{Ca}$ is the pump molecule bound by calcium and $k_1$ and $k_2$ are rate coefficients.

We substitute $z_{Ca} = 2$ and the flux of the "Ca release" reaction into Eq. (6) to obtain the pump current:

$$I_{Ca}^{pump} = k_2 \frac{2Fv}{a}[P \cdot Ca] = k_2 \frac{2Fv}{a}([P]_0 - [P]) \ , \tag{8}$$

where we have used the fact that the total concentration of the pump molecule $[P]_0$ is the sum of the concentrations $[P]$ and $[P \cdot Ca]$ of unbound and bound pump molecule. Since there is only a single compartment, we have dropped subscripts. The calcium channel current $I_{Ca}^{chan}$ flowing into the compartment is determined by Eq. (2) with a constant conductance of magnitude $g_{Ca}$, and the nonspecific current is used for the passive channels so that $I_{ns}^{chan} = g_{pas}(V - E_{pas})$, where $E_{pas}$ is the passive reversal potential.

To construct the ODEs corresponding the kinetic scheme (7) and the expression for the pump current (8), Eq. (5) is applied to the scheme to give fluxes, which, along with $I_{Ca}^{chan}$, $I_{Ca}^{pump}$ and $I_{ns}^{chan}$, are substituted in Eqs. (1) and (3) to give:

$$C \frac{dV}{dt} = -g_{Ca} f_{Ca}(V, [Ca], [Ca]_o) - k_2 \frac{2Fv}{a}([P]_0 - [P]) - g_{pas}(V - E_{pas}) \tag{9}$$

$$\frac{d[Ca]}{dt} = -\frac{a}{2Fv} g_{Ca} f(V, [Ca], [Ca]_o) - k_1[Ca][P] \tag{10}$$

$$\frac{d[P]}{dt} = -k_1[Ca][P] + k_2([P]_0 - [P]) \ . \tag{11}$$

The notional volume $v$ may describe the volume of a thin submembrane shell rather than the volume of the whole compartment. We assume that $v$ is the volume of the whole cylindrical compartment so that $a/v = 4/d$, where $d$ is the diameter of the compartment.

## 2.4    Simulation of Deterministic Variables

Simulators of deterministic electrical and chemical activity in neurons, such as NEURON [3], solve the coupled ODEs by gathering the variables $V_i$, $[S]_i$ and other state variables into one state vector $x$ and solving the ODE system:

$$\frac{dx}{dt} = G(x) + b(t) \ , \tag{12}$$

where $G(x)$ is the rate of change of each state variable and $b(t)$ is a time dependent forcing input. In principle $G(x)$ depends on all variables, though the structure of compartmental models means each element of $G(x)$ depends on only a few elements of $x$. These equations can be solved by implicit Euler integration, which, although not providing the second-order accuracy of more advanced schemes, does give guarantees about numerical stability and is used by default in NEURON [18]. In implicit Euler the derivative is evaluated at $t + \Delta t$, the end of the time step:

$$\frac{x(t + \Delta t) - x(t)}{\Delta t} = G(x(t + \Delta t)) + b(t + \Delta t) \ . \tag{13}$$

Taylor expanding the right hand side of (13) in $\Delta t$ and rearranging gives the equation for one update step:

$$x(t + \Delta t) = x(t) + \left( I - \frac{\partial G}{\partial x} \Delta t \right)^{-1} (G(x(t)) + b(t)) \, \Delta t \; , \qquad (14)$$

where $\partial G / \partial x$ is the Jacobian matrix at time $t$, which can be computed numerically, or analytically for efficiency. To optimise simulation speed, the time step $\Delta t$ may vary depending on the rate of change of the variables; but whatever the value of $\Delta t$, all variables are updated simultaneously.

## 2.5    Modifications to Accommodate Rule-Based Simulation

To combine fixed-step simulation of continuous variables described by ODEs with discrete variables described by stochastic, rule-based models, we use principles akin to those used in hybrid simulations of systems of chemical reactions [13]. The state variables (elements of $x$) are partitioned into continuous variables, which are updated at fixed intervals of $\Delta t$ by an ODE solver, and discrete variables, which are updated asynchronously by the rule-based solver, as outlined in Appendix A.1. Some "bridge" variables are referred to by both solvers. The combined simulation algorithm must ensure that the two solvers are synchronised appropriately and that conversions between continuous and discrete quantities are made.

For simulations combining molecular and electrical activity (e.g. Fig. 1) the membrane potential would be a continuous variable, intracellular molecules such as calmodulin and CaMKII would be stochastic variables, and the intracellular calcium in the spine would be a stochastic bridge variable.

*Conversions.* In deterministic simulations of biochemical reactions in neurons (e.g. [1]) a molecular species or ion $S$ is represented by an intensive quantity – its concentration $[S]$; whereas in stochastic simulations it is represented by an extensive quantity – the number of molecules $|S|$ in the volume $v$ in which $S$ exists. Thus to compare the deterministic (ODE) and stochastic (rule-based) parts of the simulation, intensive and extensive quantities need to be interconverted using Avogadro's constant $N_A$:

$$|S| = N_A v [S] \; . \qquad (15)$$

Rate coefficients for reactions based on concentrations must also be converted to ones appropriate for species number for use in the rule-based simulator's rules. To derive the conversion formula, consider a bimolecular kinetic scheme in which $k$ is the rate coefficient, i.e. $S + T \xrightarrow{k} S \cdot T$. Typical units for $k$ are $M^{-1}s^{-1}$. The kinetic scheme can be expressed as an ODE $d[S \cdot T]/dt = k[S][T]$. Converting the concentrations according to (15) yields an equivalent ODE whose variables are numbers of molecules: $d|S \cdot T|/dt = \gamma |S||T|$, where $\gamma = k/N_A v$ is the converted rate coefficient and has units $s^{-1}$. In general for an equation with $n$ reactants, the

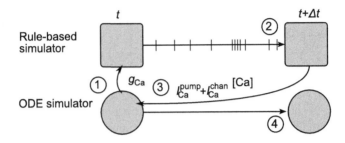

**Fig. 2.** The update and synchronisation method. See text for explanation.

relation between rate coefficients for numbers of molecules and concentrations is $\gamma = (N_A v)^{-n+1} k$. State variables (for example the state of channels) may also be controlled by the rule-based simulator, and here other conversion formulae apply.

*Creation and Destruction Rules for Bridging Variables.* To simulate the channel currents in the rule-based solver we need to write creation or destruction rules that are equivalent to $I^{chan}_{S,i}$ in Eq. (3). These rules are:

$$\xrightarrow{-a_i \tilde{I}^{chan}_{S,i} N_A / z_S F} S \qquad \text{if } \tilde{I}^{chan}_{S,i} < 0 \tag{16}$$

$$S \xrightarrow{a_i \tilde{I}^{chan}_{S,i} N_A / z_S F} \qquad \text{if } \tilde{I}^{chan}_{S,i} > 0. \tag{17}$$

Here $\tilde{I}^{chan}_{S,i}$ can be an expression that references continuous or discrete variables.

*Update and Synchronisation.* The procedure for updating the time from $t$ to $t + \Delta t$ (Fig. 2) is:

1. Pass all relevant continuous variables, e.g. conductances and voltages needed to compute $I^{chan}_{S,i}$ in the rule-based simulator.
2. Run the rule-based simulator from $t$ to $t + \Delta t$.
3. Compute the net change $\Delta S^{tot}_i$ in the total number of each bridging species $S$ (including in any complexes) in compartment $i$ over the time step and convert each change back to a current: $I^{chan}_{S,i} + I^{pump}_{S,i} = -\Delta S^{tot}_i z_S F / a_i N_A$. For each membrane potential $V_i$, set the corresponding element of $\boldsymbol{b}(t)$ equal to $-(1/C_i) \sum_S (I^{chan}_{S,i} + I^{pump}_{S,i})$ (see second term on right of Eq. 1).
4. Update the continuous variables according to the update step (14).

When running the rule-based model, it will not stop precisely on the boundary of the time step since the update times are generated stochastically. To deal with this problem, the time of the next event in the rule-based component is computed before updating the variables. As soon as the next event time is after the end of the deterministic step, that event is thrown away, as justified in Appendix A.2.

# 3   Implementation

We have implemented the algorithm described in the previous section by linking the Java-based SpatialKappa implementation of the Kappa language [24] to version 7.4 of NEURON, which allows reaction-diffusion equations to be specified in Python [18]. Our implementation ("KappaNEURON") is available at http:// github.com/davidcsterratt/KappaNEURON. We have used the py4j[1] package to extend the SpatialKappa simulator so its Java objects can be accessed in Python. The wrapper system in NEURON 7.4 allows us to override NEURON's built-in

```
## File caPump.ka - Simple calcium pump

## Agent declarations, showing the agent names and binding sites
%agent: ca(x)    # Calcium with binding site
%agent: P(x)     # Pump molecule with binding site

## Variable declarations
%var: 'vol' 1             # Volume in um3
%var: 'NA'  6.02205E23 # Avagadro's constant
# Concentration of one agent in the volume in mM
%var: 'agconc' 1E18/('NA' * 'vol')
# Rate constants in /mM-ms or /ms, depending on the number of
# complexes on LHS of rule
%var: 'k1' 0.001       # /mM-ms
%var: 'k2' 1           # /ms

## Rules
# Note the scaling of the rate constant of the bimolecular reaction
'ca binding' ca(x), P(x)     -> ca(x!1), P(x!1) @ 'k1' * 'agconc'
'ca release' ca(x!1), P(x!1) -> P(x)            @ 'k2'

## Initialisation of agent numbers
# Overwritten by NEURON but needed for SpatialKappa parser
%init:  1000 ca(x)
%init: 10000 P(x)

## Observations
%obs: 'ca'    ca(x)          # Free Ca
%obs: 'P-Ca' ca(x!1), P(x!1) # Bound Ca-P
%obs: 'P'    P(x)            # Free P
```

**Fig. 3.** Example of a Kappa file for a simple calcium pump (7) simulated in a volume of $1\,\mu m^3$. Note the conversion of the forward rate coefficient from units of $mM^{-1}\,ms^{-1}$ to $ms^{-1}$. For an introduction to the Kappa language, see the short description at http:// www.kappalanguage.org/syntax.html.

---

[1] http://py4j.sourceforge.net/.

fixed solve callback function with one that calls the SpatialKappa simulator at each time step, as described in the previous section.

In order to specify a model the Kappa component is specified in a separate file. Figure 3 shows an example of a simple calcium pump specified in Kappa. This file is then linked into the NEURON simulation as demonstrated in the Python code in Fig. 4. The mechanisms specified in the Kappa file take over all of NEURON's handling of molecules in the cytoplasm of chosen sections.

```python
from neuron import *
import KappaNEURON

## Create a compartment
sh = h.Section()
sh.insert('pas')                    # Passive channel
sh.insert('capulse')               # Code to give Ca pulse
# This setting of parameters gives a calcium influx and pump
# activation that is scale-independent
sh.gcalbar_capulse = gcalbar*sh.diam

## Define region where the dynamics will occur ('i' means intracellular)
r = rxd.Region([sh], nrn_region='i')

## Define the species, the ca ion (already built-in to NEURON), and the
## pump molecule. These names must correspond to the agent names in
## the Kappa file.
ca = rxd.Species(r, name='ca', charge=2, initial=0.0)
P  = rxd.Species(r, name='P',  charge=0, initial=0.2)

## Create the link between the Kappa model and the species just defined
kappa = KappaNEURON.Kappa(membrane_species=[ca], species=[P],
                          kappa_file='caPump.ka', regions=[r])

## Transfer variable settings to the kappa model
vol = sh.L*numpy.pi*(sh.diam/2)**2
kappa.setVariable('k1',  47.3)
kappa.setVariable('k2',  gamma2)
kappa.setVariable('vol', vol)

## Run
init()
run(30)
```

**Fig. 4.** Extract of Python code to link the Kappa file shown in Fig. 3 into a compartment in NEURON.

## 4    Results

### 4.1    Validation

We validated our implementation by comparing the results of simulating ODE and rule-based versions of the model described in Sect. 2.3 using standard NEURON and KappaNEURON respectively. Figure 5A shows the deterministic solution of the system of ODEs (9)–(11) (blue) and a sample rule-based solution using the Kappa rules in Fig. 3 (red) from a single compartment with diameter $1\,\mu m$, giving a volume within the range $0.01$–$1\,\mu m^3$ typical of spine heads in the vertebrate central nervous system [11]. The calcium conductance $g_{Ca}$ is zero apart from during a pulse lasting from 5–10 ms when an inward calcium current begins to flow ($I_{Ca}^{chan}$ is negative). This causes a sharp rise in intracellular calcium concentration [Ca], which, because of the GHK current equation, reduces the calcium current slightly, accounting for the initially larger magnitude of the calcium current. As the calcium concentration increases, it starts binding to the pump molecules, depleting the amount of the free pump molecules [P]. Once the calcium channels close ($g_{Ca} = 0$), the calcium influx stops, and the remaining free calcium is taken up quickly by the pumps. The pump-calcium complex dissociates at a slower rate, leading to a positive (outwards) calcium current. The stochastic traces (red) are very similar to their deterministic counterparts (blue) apart from some random fluctuations, particularly in the trace of calcium. This agreement, along with a suite of simpler tests included with the source code, indicates that the implementation is correct.

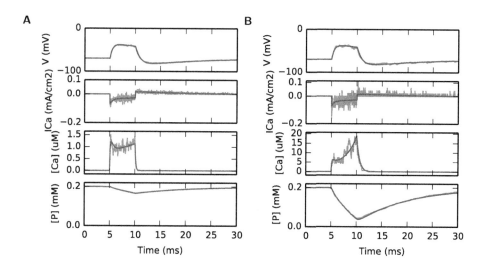

**Fig. 5.** Reference simulations. **A,** Traces generated by NEURON with NMODL (blue) and KappaNEURON (red) when the diameter is $1\,\mu m$. **B,** The same simulations but with a diameter of $0.2\,\mu m$. There is more noise evident in the combined simulation due to the smaller number of ions involved (Color figure online).

Figure 5B shows deterministic (blue) and stochastic (red) simulations in a spine with a diameter of $0.2\,\mu m$ (i.e. 1/25 of the volume of the simulation in Fig. 5A). The shape of the traces differs due to the change in surface area to volume ratio. Due to the smaller volume and hence smaller numbers of ions involved, the fluctuations are relatively bigger.

## 4.2   Demonstration Simulation

To demonstrate the utility of integrated electrical and rule-based neuronal models, we constructed a model of a subset of the synaptic proteome, with a focus on the signal processing at the early stages of the CaM-CaMKII pathway. We encoded in Kappa published models of: the dynamics of NMDARs [27]; binding of calcium with calmodulin and binding of calmodulin-calcium complexes to CaMKII [19]; and binding of calcium to a calbindin buffer [8]. We embedded these linked models into a simple model of a synaptic spine, comprising head and neck compartments, connected to a dendrite. As well as the NMDARs, modelled in Kappa, there were AMPARs in the spine head, and backpropagating spikes were modelled by inserting standard Hodgkin-Huxley ion channels [12] in the dendritic membrane. To emulate spike-timing dependent synaptic plasticity protocols [15], a train of 10 excitatory postsynaptic potentials (EPSPs) were induced in the synapse in the spine head at 20 Hz, each of which was followed by

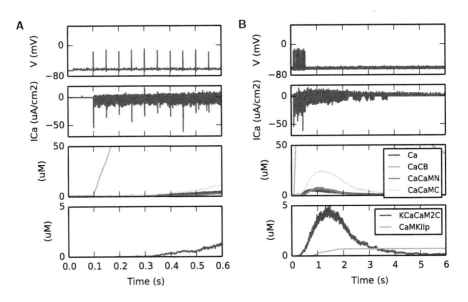

**Fig. 6.** Demonstration simulations. **A,** The first 600 ms of the simulation. **B,** The first 6000 ms of the simulation. Labels: Ca, free calcium; CaCB, calcium bound to calbindin; CaCaMN, calcium bound to the N lobe of calmodulin; CaCaMC, calcium bound to the C lobe of calmodulin; KCaCaM2C, CaMKII bound to calmodulin with two calcium ions on its C lobe; and CaMKIIp, phosphorylated CaMKII.

an action potential. There were also 50 other synaptic inputs onto the dendrite, though these did not contain the rule-based model.

Figure 6 shows results from one simulation at short (0.6 s) and long (6 s) time scales. The first stimulation of the detailed synapse paired with action potential initiation occurs at 0.1 s, as can be seen by the voltage trace. The stimulation releases glutamate, which binds to AMPARs (which remain open for a few milliseconds) and NMDARs (which remain open for 100 s of milliseconds). Due to the backpropagating action potentials releasing the voltage-dependent block of NMDARs, there are peaks in the calcium current ($I_{Ca}$) at the same time as the action potentials. The calcium entering through the NMDARs binds to the calbindin and calmodulin (CaM) buffers. The CaM-Ca$^{2+}$ complex can bind to CaMKII, the rate depending on which of the C and N lobes of CaM the Ca$^{2+}$ are bound to. This CaMKII-CaM-Ca$^{2+}$ complex can then be phosphorylated, leading to a long lasting elevation in its level. This will then to phosphorylate stargazin, which will help to anchor AMPARs in the membrane, thus contributing to LTP.

## 5   Discussion

We have presented a method for integrating a stochastically simulated rule-based model of proteins in a micron-sized region of a neuron into a compartmental model of electrical activity in the whole neuron. The rule-based component allows the biochemical interactions between binding sites on proteins to be specified using a tractable number of rules, with the simulator taking on the work of tracking which complexes are present at any point during the simulation.

Our approach is similar to that of Kiehl et al. [13] who simulated chemical reactions with a hybrid scheme. However their integration scheme synchronised at every discrete event, whereas in ours, synchronisation is driven only by the time step of the continuous simulator. This principle is appropriate for neural systems, in which we can expect many discrete events per time step.

Recently Mattioni and Le Novère [17] have integrated the ECell simulator with NEURON. Our approach is similar to theirs, though with two differences. Firstly we have integrated a rule-based simulator. This has the advantage that the interactions between the combinatorially large numbers of complexes present in the PSD can be specified using a tractable number of rules, though this limits the simulation method to the sequential style of Gillespie's stochastic simulation algorithm, and does not allow for any of the approximations that increase the algorithm's efficiency. Secondly Mattioni and Le Novère get the ODE-based solver to handle calcium, whereas we handle it in the rule-based solver. Our approach is less efficient computationally, but it ensures that all biochemical quantities are consistent and avoids having to make any assumptions about the relative speeds of processes.

Our approach allows us to model at a considerable level of detail. For example the conformation of NMDARs may be part of the biochemical model, allowing proteins in the PSD (e.g. calmodulin bound to calcium) to modulate the state of the channel [27]. We can also use one rule-based scheme to model both the presynapse and the postsynapse, which could help to understand transynaptic signalling via molecules such as endocannabinoids [4].

Our implementation of our algorithm is publicly available (KappaNEU-RON; http://github.com/davidcsterratt/KappaNEURON) and under development. The next major feature planned is making available to NEURON SpatialKappa's capability of simulating rule-based models with voxel-based diffusion.

# A    Appendix

## A.1    Kappa Simulation Method

To understand the asynchronous nature of the Kappa simulation method, we first illustrate Gillespie's direct method [10] by applying it to the kinetic scheme description of a calcium pump shown in Eq. (7). Here Ca represents intracellular calcium, P represents a pump molecule in the membrane, $P \cdot Ca$ is the pump molecule bound by calcium and $k_1$ and $k_2$ are rate coefficients, which are rescaled to the variables $\gamma_1$ and $\gamma_2$ as explained in Sect. 2.5. To apply the Gillespie method to this scheme:

1. Compute the *propensities* of the reactions $a_1 = \gamma_1 |Ca||P|$ and $a_2 = \gamma_2 |Ca \cdot P|$
2. The total propensity is $A = a_1 + a_2$
3. Pick reaction $R_i$ with probability $a_i/A$
4. Pick time to reaction $T = -(\ln r)/A$, where $r$ is a random number drawn uniformly from the interval $(0, 1)$.
5. Goto 1

Kappa uses an analogous method, but applied to rules that are currently active. Both methods are event-based rather than time-step based.

## A.2    Justification for Throwing Away Events

To justify throwing away events occurring after a time step ending at $t + \Delta t$, we need to show that the distribution of event times (measured from $t$) is the same in two cases:

1. The event time $T$ is drawn from an exponential distribution $A \exp(-AT)$ (for $T > 0$), where $A$ is the propensity.
2. An event time $T_0$ is drawn as above. If $T_0 < \Delta t$, accept $T = T_0$ as the event time. If $T_0 \geq \Delta t$, throw away this event time and sample a new interval $T_1$ from an exponential distribution with a time constant of $A$, i.e. $A \exp(-AT_1)$. Set the event time to $T = \Delta t + T_1$.

In the second case, the overall distribution is:

$$
\begin{aligned}
P(\text{event at } T < \Delta t) &= A \exp(-AT) \\
P(\text{event at } T \geq \Delta t) &= P(\text{survival to } \Delta t)P(\text{event at } T_1) \\
&= \exp(-A\Delta t)A \exp(-A(T - \Delta t)) \\
&= A \exp(-AT)
\end{aligned}
\tag{18}
$$

Here we have used $T_1 = T - \Delta t$. Thus the distributions are the same in both cases.

# References

1. Bhalla, U.S., Iyengar, R.: Emergent properties of networks of biological signalling pathways. Science **283**, 381–387 (1999)
2. Cannon, R.C., O'Donnell, C., Nolan, M.F.: Stochastic ion channel gating in dendritic neurons: morphology dependence and probabilistic synaptic activation of dendritic spikes. PLoS Comput. Biol. **68**, e1000886 (2010)
3. Carnevale, T., Hines, M.: The NEURON Book. Cambridge University Press, Cambridge (2006)
4. Castillo, P.E., Younts, T.J., Chávez, A.E., Hashimotodani, Y.: Endocannabinoid signaling and synaptic function. Neuron **761**, 70–81 (2012)
5. Chylek, L.A., Stites, E.C., Posner, R.G., Hlavacek, W.S.: Innovations of the rule-based modeling approach. In: Prokop, A., Csukás, B. (eds.) Systems Biology, pp. 273–300. Springer, Heidelberg (2013)
6. Collins, M.O., Husi, H., Yu, L., Brandon, J.M., Anderson, C.N.G., Blackstock, W.P., Choudhary, J.S., Grant, S.G.N.: Molecular characterization and comparison of the components and multiprotein complexes in the postsynaptic proteome. J. Neurochem. **97**, 16–23 (2006)
7. Danos, V., Feret, J., Fontana, W., Krivine, J.: Scalable simulation of cellular signaling networks. In: Shao, Z. (ed.) APLAS 2007. LNCS, vol. 4807, pp. 139–157. Springer, Heidelberg (2007)
8. Faas, G.C., Raghavachari, S., Lisman, J.E., Mody, I.: Calmodulin as a direct detector of $Ca^{2+}$ signals. Nat. Neurosci. **143**, 301–304 (2011)
9. Faeder, J., Blinov, M., Hlavacek, W.: Rule-based modeling of biochemical systems with BioNetGen. In: Maly, I.V. (ed.) Systems Biology, Methods in Molecular Biology, vol. 500, pp. 113–167. Humana Press (2009)
10. Gillespie, D.: Exact stochastic simulation of coupled chemical reactions. J. Phys. Chem. **81**, 2340–2361 (1977)
11. Harris, K.M., Kater, S.B.: Dendritic spines: cellular specializations imparting both stability and flexibility to synaptic function. Annu. Rev. Neurosci. **17**, 341–371 (1994)
12. Hodgkin, A.L., Huxley, A.F.: A quantitative description of membrane current and its application to conduction and excitation in nerve. J. Physiol. (Lond.) **117**, 500–544 (1952)
13. Kiehl, T.R., Mattheyses, R.M., Simmons, M.K.: Hybrid simulation of cellular behavior. Bioinformatics **203**, 316–322 (2004)
14. Lisman, J.E., Zhabotinsky, A.M.: A model of synaptic memory: a CaMKII/PP1 switch that potentiates transmission by organizing an AMPA receptor anchoring assembly. Neuron **312**, 191–201 (2001)

15. Markram, H., Lübke, J., Frotscher, M., Sakmann, B.: Regulation of synaptic efficiency by coincidence of postsynaptic APs and EPSPs. Science **275**, 213–215 (1997)
16. Martin, S.J., Grimwood, P.D., Morris, R.G.M.: Synaptic plasticity and memory: an evaluation of the hypothesis. Annu. Rev. Neurosci. **231**, 649–711 (2000)
17. Mattioni, M., Le Novère, N.: Integration of biochemical and electrical Signaling-Multiscale model of the medium spiny neuron of the striatum. PLoS ONE **87**, e66811 (2013)
18. McDougal, R.A., Hines, M.L., Lytton, W.W.: Reaction-diffusion in the NEURON simulator. Front. Neuroinform. **7**, 1–13 (2013)
19. Pepke, S., Kinzer-Ursem, T., Mihalas, S., Kennedy, M.B.: A dynamic model of interactions of $Ca^{2+}$, calmodulin, and catalytic subunits of $Ca^{2+}$/calmodulin-dependent protein kinase II. PLoS Comput. Biol. **62**, e1000675 (2010)
20. Pocklington, A.J., Cumiskey, M., Armstrong, J.D., Grant, S.G.N.: The proteomes of neurotransmitter receptor complexes form modular networks with distributed functionality underlying plasticity and behaviour. Mol. Syst. Biol. **2**, 1–14 (2006)
21. Ray, S., Bhalla, U.S.: PyMOOSE: interoperable scripting in python for MOOSE. Front. Neuroinform. vol. 2(6) (2008)
22. Smolen, P., Baxter, D.A., Byrne, J.H.: A model of the roles of essential kinases in the induction and expression of late long-term potentiation. Biophys. J. **908**, 2760–2775 (2006)
23. Sorokina, O., Sorokin, A., Armstrong, J.D.: Towards a quantitative model of the post-synaptic proteome. Mol. Biosyst. **7**, 2813–2823 (2011)
24. Sorokina, O., Sorokin, A., Armstrong, J.D., Danos, V.: A simulator for spatially extended kappa models. Bioinformatics **29**, 3105–3106 (2013)
25. Sterratt, D., Graham, B., Gillies, A., Willshaw, D.: Principles of Computational Modelling in Neuroscience. Cambridge University Press, Cambridge (2011)
26. Stiles, J.R., Bartol, T.M.: Monte Carlo methods for simulating realistic synaptic microphysiology using MCell. In: De Schutter, E. (ed.) Computational Neuroscience: Realistic Modeling for Experimentalists, Chap. 4, pp. 87–127. CRC Press, Boca Raton (2001)
27. Urakubo, H., Honda, M., Froemke, R.C., Kuroda, S.: Requirement of an allosteric kinetics of NMDA receptors for spike timing-dependent plasticity. J. Neurosci. **2813**, 3310–3323 (2008)
28. Zeng, S., Holmes, W.R.: The effect of noise on CaMKII activation in a dendritic spine during LTP induction. J. Neurophysiol. **1034**, 1798–1808 (2010)

# FM-Sim: A Hybrid Protocol Simulator of Fluorescence Microscopy Neuroscience Assays with Integrated Bayesian Inference

Donal Stewart[1]($\boxtimes$), Stephen Gilmore[2], and Michael A. Cousin[3]

[1] Doctoral Training Centre in Neuroinformatics and Computational Neuroscience,
School of Informatics, University of Edinburgh, Edinburgh, UK
donal.stewart@ed.ac.uk

[2] Laboratory for Foundations of Computer Science, School of Informatics,
University of Edinburgh, Edinburgh, UK
stephen.gilmore@ed.ac.uk

[3] Centre for Integrative Physiology, School of Biomedical Sciences,
University of Edinburgh, Edinburgh, UK
m.cousin@ed.ac.uk

**Abstract.** We present FM-Sim, a domain-specific simulator for defining and simulating fluorescence microscopy assays. Experimental protocols as performed *in vitro* may be defined in the simulator. The defined protocols then interact with a computational model of presynaptic behaviour in rodent central nervous system neurons, allowing simulation of fluorescent responses to varying stimuli. Rate parameters of the model may be obtained using Bayesian inference functions integrated into the simulator, given experimental fluorescence observations of the protocol performed *in vitro* as training data. These trained protocols allow for predictive *in silico* modelling of potential experimental outcomes prior to time-consuming and expensive *in vitro* studies.

## 1 Introduction

Synaptic vesicle recycling at the pre-synaptic terminal of neurons is essential for the maintenance of neurotransmission at central synapses. Among the tools used to visualise the mechanics of this process is time-series fluorescence microscopy. Fluorescent dyes such as FM1-43, or engineered fluorescent versions of synaptic vesicle proteins such as pHluorins, have been employed to reveal different steps of this key process. These tools have been applied to a number of animal models, notably the neurons within the central nervous system in rodents.

Predictive *in silico* modelling of potential experimental outcomes is a highly informative procedural step prior to making the significant investment of time and expense which is needed to prepare and run an *in vitro* study. FM-Sim is a hybrid stochastic simulator written specifically for this domain. It allows the definition of many of the experimental assays which are commonly used in this field, and simulates them against a computational model of the current understanding of the mechanisms of synaptic vesicle recycling.

© Springer International Publishing Switzerland 2015
O. Maler et al. (Eds.): HSB 2013 and 2014, LNBI 7699, pp. 159–174, 2015.
DOI: 10.1007/978-3-319-27656-4_10

If experimental observation data is available from *in vitro* fluorescence microscopy assays, then simulated time-series output from the computational model may be checked for its distance from the observations, and rate parameters can be automatically inferred to fit the provided observations. FM-Sim will provide high predictive power in studies examining presynaptic function.

The main benefit of simulations performed by FM-Sim over those performed by generic stochastic simulators is that FM-Sim manages the effects of nested and overlapping events which impact on the kinetics of vesicle recycling. For example, a chemical inhibitor may be introduced at some time during an experimental protocol, with electrical stimulation applied subsequently for part of the time that the inhibitor is available. Additionally, FM-Sim can simulate a number of *in vitro* techniques used to modify the fluorescence output during an experiment.

## 2    Biological Background

The domain of interest to this study is the presynaptic terminal of central nervous system neurons, as found in experimental animal models such as rodents. This is an active domain of experimental study, where the research goal is to discover the mechanisms which control the processing of synaptic vesicles within chemical synapses.

### 2.1    The Synaptic Vesicle Cycle

Within chemical synapses of central nervous system (CNS) neurons, neurotransmitter is released from the presynaptic terminal to propagate the neural signal to the postsynaptic terminal of the following neuron. This neurotransmitter is stored in vesicles within the presynaptic terminal, which are exocytosed in response to an incoming action potential (Fig. 1). These vesicles are classed as being within pools denoting their availability for release via exocytosis: the *readily releasable pool* consists of vesicles ready to be released immediately, while the *reserve pool* consists of vesicles filled with neurotransmitter, but not close enough to the plasma membrane for immediate release.

To prevent depletion of these vesicle pools, compensatory endocytosis of plasma membrane allows regeneration of these vesicles. Two forms are studied within CNS nerve terminals:

- **Clathrin Mediated Endocytosis (CME)** [16] is the primary mechanism of membrane recovery during periods of normal levels of nerve stimulation. Individual vesicles are reconstructed directly from the plasma membrane. Following reacidification of the vesicle contents and refilling with neurotransmitter, these vesicles rejoin the reserve and readily releasable pools.
- **Activity Dependent Bulk Endocytosis (ADBE)** [9] is a second endocytosis mechanism triggered by periods of high stimulation. Here, large areas of plasma membrane are endocytosed as endosomes, which are later broken down into individual vesicles for reuse.

The mechanisms of both of these forms of endocytosis are currently subjects of detailed study.

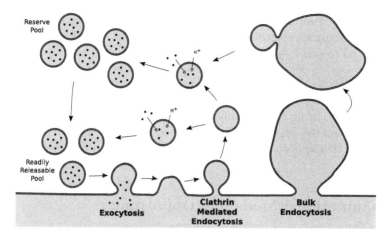

**Fig. 1.** The synaptic vesicle cycle.

## 2.2 Fluorescence Microscopy Imaging

Time-series fluorescence microscopy is one of the experimental procedures used to study the mechanisms of the synaptic vesicle cycle. Fluorescent probes added to nerve terminals allow us to obtain time-series images of nerve terminal behaviour under stimulation. The two commonly-used forms of fluorescent probes are FM dyes, and engineered fluorescent synaptic vesicle proteins such as pHluorins.

- **FM dyes** [8] are soluble dyes added to the extracellular media. They fluoresce when bound to membrane. If a nerve terminal is stimulated when the dye is present in the extracellular media around the neuron, the dye is taken up within vesicles and endosomes during endocytosis. Subsequent washing of the neuron removes all extracellular dye, leaving only the dye within the nerve terminal structures to fluoresce. During exocytosis, this dye is released back into the extracellular media and removed from view, which allows tracking of exocytosis rates.

  Different FM dyes have different uptake rates depending on the type of endocytosis. This allows isolation of the behaviour of the different forms of endocytosis. Two commonly-used dyes are FM1-43 and FM2-10.
- **pHluorins** [17,18] are synaptic vesicle proteins fused to a fluorescent protein. These fluoresce only at neutral pH $\approx$ 7.4 (as found in the extracellular media), and do not fluoresce at low pH levels $\approx$ 5.5 (as found in reacidified vesicles). Over time the vesicles are endocytosed and reacidified causing a change in the level of fluorescence which can be used to track activity in the synaptic vesicle cycle.

  Images taken during an experimental assay show many terminals of a neuron simultaneously. At the imaging resolutions which are commonly used, an individual synaptic terminal appears as a single bright area a few pixels in diameter and thus only the fluorescence of the terminal as a whole can be measured. Ingenuity in experimental design is required to deconstruct this single measure per nerve terminal to highlight the mechanisms of the synaptic vesicle cycle.

Different aspects of the synaptic vesicle cycle can be isolated and studied by using either FM dyes or pHluorins in combination with chemical inhibitors, or on various genetic knockdown animal models.

For example, adding the reacidification inhibitor Bafilomycin A1 to a pHluorin assay prevents the quenching of the fluorescent marker following endocytosis [3], offering further opportunities to isolate parts of the synaptic vesicle recycling process.

## 3    Computational Modelling Opportunity

This domain has already been the subject of computational modelling research. Prior work has included a study by Atluri and Ryan [3] fitting ODEs to experimental results of vesicle reacidification. This has yielded a detailed kinetic model for the reacidification step of vesicle recycling. A paper by Gabriel *et al.* [11] used experimental data showing vesicle release depression over extended stimulation to iteratively construct an ODE model representing the rate limiting steps in vesicle trafficking. A study by Granseth and Legnado [14] fitted a system of ODEs representing a model of vesicle recycling to experimental data observed in cultured hippocampal neurons. The model illustrated the effects of ambient versus physiological temperature on recycling kinetics.

Although many general-purpose simulation and analysis packages exist, most do not allow any alteration to the system during the course of an experimental assay. These alterations can include the addition and removal of reagents to extracellular media, and electrical stimulation of neurons being observed. None give a treatment of fluorescence which reflects our current understanding.

Debate is ongoing [2,3] over which types of endocytosis may be in effect at CNS nerve terminals under different stimuli. A number of published computational models investigate the ability of one or a combination of these models to fit observed experimental data [14,15]. However, these models are created on an ad-hoc basis for particular experimental protocols.

Additionally, different fluorescent markers have different properties and behaviours within these experimental assays. Generally, assays using one or other of these markers are modelled independently.

## 4    FM-Sim

We have developed FM-Sim with the intention of creating an easy-to-use application for the definition and simulation of fluorescence microscopy experiments. It is composed of a protocol definition user interface, a stochastic simulation engine, and a Bayesian inference engine to infer rate parameters for a protocol based on comparison with supplied observed data.

The simulation and inference engines use a model of vesicle movement within a nerve terminal, and can process the changes in rate which occur with changes in stimulus during a protocol, termed regime changes for the remainder of this paper. These regime changes are derived from the protocol definition, managing the effects of nested and overlapping events.

The flexibility of the protocol definition allows the wide variety of potential experiments to be modelled. New experiments can be simulated based upon rate parameters obtained from prior similar experiments.

## 4.1   FM-Sim Synaptic Terminal Model

At the core of FM-Sim is a stochastic model of vesicle movement, or more specifically the cell membrane making up the surface of vesicles, within a synaptic terminal. The model tracks the movement of cell membrane around the structures of a single synaptic terminal, namely the vesicles, endosomes and the plasma membrane. As a simplification, all membrane movements are tracked in quanta of vesicles as this is a consistent, smallest unit of membrane surface area to move. The plasma membrane is a repository of $x$ vesicles worth of membrane, and endosomes are created with $y$ vesicles worth of membrane and decrease in size as vesicles bud from their surface. Vesicles stored within the synaptic nerve terminal are designated as being part of the reserve pool (RP) or readily releasable pool (RRP).

The initiation of movement of vesicles between the regions (RP, RRP, endosomes, plasma membrane) is stochastic, with propensity defined by rules for each transition, as depicted in Fig. 2.

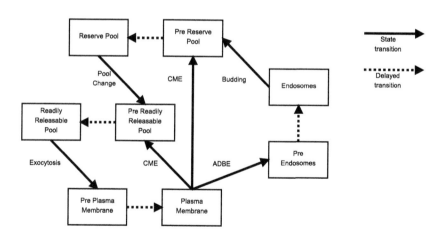

**Fig. 2.** FM-Sim synaptic terminal model.

However, in contrast to the widely-used convention in the simulation of stochastic chemical kinetics where the reaction events are abstracted as instantaneous, each of the processes moving vesicles from one state to another takes a non-negligible time. Therefore the model is extended to include delays. Each pool or region of vesicles has an associated transition state. Vesicles in this transition state take a fixed time to transition from their prior to their current state. This approach is a form of the Delayed Stochastic Simulation Algorithm (DSSA) as discussed in [4], with reactants being vesicles in the source region, and products being the vesicles transferred to the target region.

Let $S$ be the set of states in which a vesicle can exist

$$S = \{\text{RP, RRP, E, PM, Pre-RP, Pre-RRP, Pre-E, Pre-PM}\}$$

denoting Reserve Pool, Readily Releasable Pool, Endosome, Plasma Membrane, and the associated pre-states respectively.

Define a vesicle as $V_s(D, E, pH, T)$, $s \in S$ with the following properties:

– $D$ - Presence of FM dye: This determines whether or not the vesicle counts towards the fluorescence level of the nerve terminal. It is treated as a boolean value, as is the convention in current biological experiments. The dye is either present at the concentrations used when first loaded into the terminal $(D_Y)$, or completely absent $(D_N)$. Partial or low concentration dye loading does not take place.

– $E$ - Endocytosis source: The vesicle is classified as having been produced via recycling by CME $(E_{CME})$ or ADBE $(E_{ADBE})$. While it is possible that new vesicles could arrive from the soma of the cell, or be trafficked from neighbouring nerve terminals [10], it is thought that the effect of this possible influx of new vesicles occurs too slowly over the timeframe of an experiment to be detectable. The mechanism of vesicle creation is a factor in determining which vesicles fluoresce under the effect of each FM dye. FM2-10 has been shown to be taken up through CME only, whereas FM1-43 is taken up through both CME and ADBE at the loading concentrations commonly used in the literature. This differential takeup of different dyes was explored experimentally in depth in [7].

– $pH$: Extracellular media is normally of neutral pH $(pH_{neutral}$, normalised to 1). Vesicles competent for release are at lower pH $(pH_{acid}$, normalised to 0). The reacidification process occurs between endocytosis and arrival at the vesicle pool. Experiments using pHluorins are affected by this pH change. Any pHluorins in contact with neutral pH fluoresce strongly, and this fluorescence decreases as the media within the vesicle is acidified. As experimental work has provided evidence that this reacidification process can take between three and five seconds [3,15], pH levels are recorded as continuous values. After endocytosis, these values are reduced linearly over time to reach the target pH.

– $T$ - Time delay for vesicle to remain in pre-pool state as described above, with $T_0$ representing 0 time delay.

Note: for notational convenience, vesicle properties irrelevant to particular transitions may be omitted, for example $V_s(D_Y) = V_s(D_Y, E_?, pH_?, T_?)$.

## 4.2   Discrete State Changes

The state changes of the model are:

$$\text{Pool change}: \qquad V_{\text{RP}}(T_0) \xrightarrow{k_1} V_{\text{Pre-RRP}}(T_{\text{Pool change}})$$

$$\text{Exocytosis}: \quad V_{\text{RRP}}(D_?, pH_?, T_0) \xrightarrow{k_2} V_{\text{Pre-PM}}(D^*, pH_{\text{neutral}}, T_{\text{Exocytosis}})$$

$$\text{CME}: \qquad V_{\text{PM}}(D^*, E_?, T_0) \xrightarrow{k_3} V_{\text{Pre-RP}}(D^*, E_{\text{CME}}, T_{\text{CME}})$$

$$: \qquad V_{\text{PM}}(D^*, E_?, T_0) \xrightarrow{k_4} V_{\text{Pre-RRP}}(D^*, E_{\text{CME}}, T_{\text{CME}})$$

$$\text{Budding}: \qquad V_{\text{E}}(T_0) \xrightarrow{k_5} V_{\text{Pre-RP}}(T_{\text{Budding}})$$

$D^*$ represents that the vesicle is FM dye tagged *iff* FM dye is present in the extracellular media. There are two CME state changes, one for each destination pool, and each with a different propensity. Each of the rules above apply to individual vesicles for each instance of the rule being executed. The rule below however, operates on $N$ vesicles taken together to represent a single endosome. If $N$ vesicles are not available in the plasma membrane state, the rule will not be executed.

$$\text{ADBE}: \quad V_{\text{PM}}(D^*, E_?, 0) \xrightarrow{k_6} V_{\text{Pre-E}}(D^*, E_{\text{ADBE}}, T_{\text{ADBE}})$$

The values $k_1, \ldots, k_6$ are the rule propensities, each the product of the number of vesicles available in the source state, and a rule rate $C$. Let $V$ denote number of vesicles:

$$k_1 = V_{\text{RP}} \times C_{\text{Pool change}}$$
$$k_2 = V_{\text{RRP}} \times C_{\text{Exocytosis}}$$
$$k_3 = V_{\text{PM}} \times C_{\text{CME}} \times C_{\text{CME RP ratio}}$$
$$k_4 = V_{\text{PM}} \times C_{\text{CME}} \times (1 - C_{\text{CME RP ratio}})$$
$$k_5 = V_{\text{E}} \times C_{\text{Budding}}$$
$$k_6 = V_{\text{PM}} \times C_{\text{ADBE}}$$

The values $k_3$ and $k_4$ have an additional input, $C_{\text{CME RP ratio}}$. This constant takes values from zero to one and represents the proportion of vesicles produced by CME which go to the RP. The remainder go to the RRP. In this work we will use 0.7 as the value for the $C_{\text{CME RP ratio}}$. Similar values are used in the literature [5].

In summary, the input parameters for the model are the rate constants $C_?$ and the time constants $T_?$.

## 4.3   Continuous State Changes

In addition to the above state change rules, there are the following continuous processes:

– Pre-pool delays: A vesicle $T_{\mathrm{delay}} > 0$ reduces as simulated time elapses. Once the $T_{\mathrm{delay}}$ reaches $T_0$, a vesicle in a pre-pool moves into the corresponding pool

$$V_{\mathrm{Pre\text{-}pool}}(T) \rightarrow V_{\mathrm{Pre\text{-}pool}}(T - dt)$$
$$V_{\mathrm{Pre\text{-}pool}}(T_0) \rightarrow V_{\mathrm{Pool}}(T_0)$$

– Reacidification: The pH of a fully internalised vesicle or endosome (i.e. one which is no longer in direct contact with extracellular media) reduces linearly over time from $pH_{\mathrm{neutral}}$ to $pH_{\mathrm{acid}}$, at a rate matching current literature [3,15].

$$V(pH) \rightarrow V(pH - dpH)$$

### 4.4  Fluorescence Calculation

The model can simulate the fluorescence of each of the three previously described fluorescent reporters. Based on the fluorescent reporter used in the experiment design, fluorescence is calculated as described below. In all cases fluorescence is scaled from minimum $= 0$ to maximum $= 1$ possible fluorescence by dividing by the total vesicle count in the model $[V]$.

– **FM1-43**: Total fluorescence is the count of all vesicles in states not in contact with extracellular media with FM dye boolean variable true. Vesicles in states {PM, Pre-PM} do not contribute because standard experimental protocols [20] have shown FM dye rapidly dissociates from the plasma membrane and is washed away in the extracellular media.

$$F_{\mathrm{FM1\text{-}43}} = [V_s(D_Y)]/[V],\ s \notin \{\mathrm{PM},\ \mathrm{Pre\text{-}PM}\}$$

– **FM2-10**: Total fluorescence is the count of all vesicles in the RP and RRP (and their pre-states) with vesicle source CME. ADBE-derived vesicles do not contribute under standard experimental protocols of $100\mu\mathrm{M}$ FM2-10 [8].

$$F_{\mathrm{FM2\text{-}10}} = [V_s(D_Y, E_{\mathrm{CME}})]/[V],\ s \in \{\mathrm{RP},\ \mathrm{RRP},\ \mathrm{Pre\text{-}RP},\ \mathrm{Pre\text{-}RRP}\}$$

– **pHluorin**: Total fluorescence is the sum of normalised vesicle pH of all vesicles. A family of pHluorin fusion proteins is commonly used in these experiments. For the purposes of simulation however, they are broadly the same in behaviour. They fluoresce when in contact with neutral pH and not when in an acidic environment. In addition to the normal physiological changes in pH within a nerve terminal, there are two external stimuli used in pHluorin assays.
  - *Ammonium addition*: Ammonium chloride is used to cause all pHluorin in a nerve terminal to fluoresce. When applied, ammonium ions permeate extracellular media, cytoplasm, and vesicle interiors, reducing the acidity of all media [19].
  - *Acid addition*: Impermanent acid is used to quench the fluorescence of pHluorin on the plasma membrane, leaving only the pHluorin on fully internalised membrane to fluoresce [3].

$$
F_{\mathrm{pHluorin}} =
\begin{cases}
1, & \text{if NH}_4^+ \text{ present} \\
\left(\sum pH\right)/[V], \quad V_s(pH),\ s \notin \{\text{PM, Pre-PM}\} & \text{if acid present} \\
\left(\sum pH\right)/[V], \quad V_s(pH),\ s \in \mathcal{S} & \text{otherwise}
\end{cases}
$$

Current experimental data available records fluorescence levels per nerve terminal as relative measures only. It is not commonly feasible to calculate absolute quantities of marker. Therefore, experimental observations are taken to vary from maximum (1) and minimum (0) fluorescence per nerve terminal during the experimental assay.

### 4.5 Protocol Definition and Rate Parameter Specification

FM-Sim allows the user to define an experimental protocol to a sufficient level of detail to be simulated. A protocol consists of a number of events with a defined start time (or frame) and a given duration. Multiple events may be active simultaneously, for example chemical stimulation of nerve terminals while simultaneously inhibiting certain protein functions within the nerve terminal with the application of chemical inhibitors. Figure 3 shows a defined example protocol.

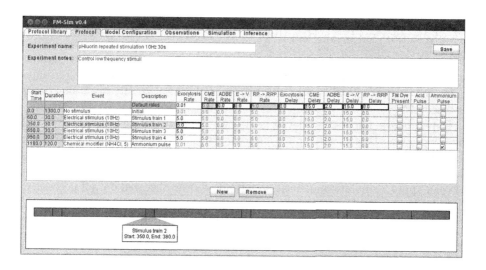

**Fig. 3.** A sample defined protocol in the FM-Sim protocol editor.

For each protocol event, the user may choose to enter values for rate parameters. The rate parameters available for modification depend on the configuration of the model to be simulated (discussed below). The user may also leave rate values for individual events unmodified, in which case the values are taken from the enclosing event if any, or a default rate if there are no enclosing events. Finally, the user may choose to mark rate parameters as to be inferred from observed data. This will require the loading of an appropriate dataset of observed fluorescence values for inference to be performed.

## 4.6   Simulation

Given a defined experimental protocol, FM-Sim calculates the time-points at which regime changes occur. These are points when the kinetic rates of the model are allowed to change, at the start or end of protocol events. For example, the onset of low intensity electrical stimulation is expected to trigger exocytosis of synaptic vesicles, and compensatory clathrin mediated endocytosis.

These regime change time-points and the associated user-defined rates in effect during each regime are used to stochastically simulate a population of nerve terminals. The result is a trace of mean simulated fluorescence during the experiment.

The simulation engine uses an extended form of the Gillespie Stochastic Simulation Algorithm, specifically, the Direct Method [12], keeping track of vesicles and endosomes within a presynaptic terminal. The extensions added allow vesicle movement to be non-instantaneous using an algorithm similar to the DSSA method by Barrio et al. [4], and handling of continuous state changes (pH) over time.

## 4.7   Observation Data Handling

Observed experimental data may be input in the form of a comma-separated value (CSV) data file holding fluorescence values across multiple nerve terminal regions-of-interest (ROIs) over multiple imaging frames (usually one frame every few seconds).

Experimental fluorescence levels are known to decay during the time course of the experiment. This is due to photobleaching of the dye during image acquisition under the microscope. It is standard practice to compensate for this decay during analysis of experimental data, most commonly with a linear decay function. FM-Sim can apply a similar compensation formula to raw observed data if necessary:

$$\text{Corrected observation} = \text{Observation} - C \times (t_{obs} - t_0)$$

where $C$ is the linear correction factor, $t_{obs}$ is the time of observation, and $t_0$ is the time at the start of image acquisition. Generally, the linear correction factor is obtained by the user using calibration assays. If more complex decay correction is required, it is expected that the user performs the preprocessing prior to importing the data into FM-Sim. Once the data has been decay-corrected and scaled, it can be compared with the output of the simulation, or used to infer rate parameters.

## 4.8    Bayesian Parameter Inference

We take a Bayesian approach to parameter inference, using a standard Particle Marginal Metropolis-Hastings (PMMH) scheme and Sequential Monte Carlo (SMC) estimates of marginal likelihoods [1,13], with the addition of the following mechanism to handle regime changes. A defined protocol is decomposed into rate change events. The minimum set of inference parameters is also calculated over the entire protocol. For each PMMH iteration, a set of proposals are drawn for the set of inference parameters. The SMC scheme is then run, changing the rates in effect as each rate change time is reached in the simulation.

The SMC scheme used in FM-Sim is as described by Golightly *et al.* [13]. It requires a set of $N$ particles generated by forward simulation for each observed time-point. The weight $w$ of each of these particles contributes to the estimated marginal likelihood $\hat{p}(y_{t+1}|\mathbf{y}_t, c)$ where $y_{t+1}$ is the experimental observation at time $t + 1$, $\mathbf{y}_t$ is the set of experimental observations up to time $t$, and $c$ is the set of simulation parameters.

$$\hat{p}(y_{t+1}|\mathbf{y}_t, c) = \frac{1}{N} \sum_{i=1}^{N} w_{t+1}^i \quad \text{where} \quad w_t^i = p(y_t|x_t^i, c)$$

Each particle weight is $p(y_t|x_t^i, c)$, where $y_t$ is the experimentally observed (normalised) fluorescence at time $t$, $x_t^i$ is the fluorescence of simulated particle $i$ at time $t$, and $c$ is the set of simulation parameters. In other words, the probability of obtaining values matching the observed $y_t$ given the simulated result $x_t$, making the assumption that $p(y_t|x_t^i, c) = p(y_t|x_t^i)$.

The calculations also make the assumption that both the experimental and simulated observations at a particular time-point are normally distributed. Experimental observations provide a set of fluorescence values for $y_t$, from which we can compute a mean $\mu(y_t)$ and variance $\sigma(y_t)^2$. Forward simulation of an individual particle provides a single value for $x_t^i$. Given that the single particle result has zero variance, $\mathcal{N}(\mu(y_t)|x_t, \sigma(x_t)^2)$ is not useful.

To avoid this problem, we make the assumption that the variance around an $x_t^i$ result is equal to the variance of the observed $y_t$ values. This has the advantage over choosing an arbitrary variance that it provides a variance for each time-point which is tailored to the experimental data. This provides the distribution

$$\mathcal{N}(\mu(y_t)|x_t^i, \sigma(y_t)^2) = \mathcal{N}(x_t^i|\mu(y_t), \sigma(y_t)^2)$$

Finally, the distribution is normalised to the peak probability $\mathcal{N}(\mu(y_t)|\mu(y_t), \sigma(y_t)^2)$ to avoid arithmetic underflow of marginal likelihoods of long time-series, and to provide a more user-friendly distance measure. As PMMH proposal acceptance probability is based on the ratio of proposal marginal likelihood over the previously accepted marginal likelihood, this normalisation of the whole time series has no effect on proposal acceptance.

$$p(y_t|x_t^i, c) = \frac{\mathcal{N}(x_t^i|\mu(y_t), \sigma(y_t)^2)}{\mathcal{N}(\mu(y_t)|\mu(y_t), \sigma(y_t)^2)}$$

$$= \exp\left(-\frac{x_t^i - \mu(y_t)}{2\sigma(y_t)^2}\right)$$

The overall estimated marginal likelihood of the SMC scheme is the product of the marginal likelihoods of all of the $T$ observed time-points. This marginal likelihood is used to update the PMMH best match for that iteration.

$$\hat{p}(\mathbf{y}|c) = \hat{p}(y_1|c) \prod_{t=1}^{T-1} \hat{p}(y_{t+1}|\mathbf{y}_t, c)$$

$$= \prod_{t=1}^{T} \frac{1}{N} \sum_{i=1}^{N} p(y_t|x_t^i, c)$$

The final result of the inference process is a set of model rate parameters for the whole protocol which best matched the experimental data, from the proposals generated by the inference algorithm. For most applications, it is expected that there will be a mix of user-defined and inferred rate parameter values.

## 5    Case Study 1: FM Dye Based Assay

This case study uses real experimental data from the Cousin neuronal cell biology laboratory in Edinburgh [5]. The assay has multiple exocytosis stimuli designed to trigger release of synaptic vesicles from both of the vesicle pools (RP and RRP). The case study demonstrates use of FM dyes, regime changes and a series of protocol events, shown in Table 1. The raw observations are corrected for decay with a linear correction of 200 fluorescence units per second.

**Table 1.** Protocol events for the FM dye based assay.

| Event | Stimulus | Start time (s) | Duration (s) |
|---|---|---|---|
| Default | None | 0 | 250 |
| RRP Unload | Electrical (20 Hz) | 12 | 2 |
| RP Unload | Chemical (KCl, 0.5 mM) | 100 | 30 |

This first example shows inferred parameters for a model without ADBE. The inference parameters were Exocytosis, CME, and Pool Change rate, for each of the three protocol events. The remaining parameters were fixed. The final results of inference are shown in Fig. 4a, leading to the parameters shown in Table 2, with a distance measure of −0.35.

Repeating the inference with allowing the parameters ADBE and Budding rate to be inferred under the second stimulus gave the parameters shown also in Table 2, with an improved distance measure of −0.22, illustrated in Fig. 4b.

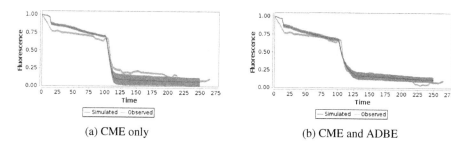

(a) CME only                          (b) CME and ADBE

**Fig. 4.** Simulated fluorescence traces using the inferred parameters shown in Table 2, plotted against the real observed data used for parameter inference.

**Table 2.** Inference results for CME only, and both CME and ADBE. Values shown are rate constants $C$ $(s^{-1})$ and time delay constants $T$ (s) for each of the discrete state change rules, with inferred values shown in bold.

| Event | Exocytosis | | CME | | ADBE | | Budding | | Pool change | |
|---|---|---|---|---|---|---|---|---|---|---|
| | $C$ | $T$ | $C$ | $T$ | $C$ | $T$ | $C$ | $T$ | $C$ | $T$ |
| CME only | | | | | | | | | | |
| Default | **0.03** | 0 | **0.01** | 15 | 0 | 2 | 0 | 15 | **0.00** | 1 |
| RRP unload | **8.81** | 0 | **1.67** | 15 | 0 | 2 | 0 | 15 | **0.92** | 1 |
| RP unload | **8.23** | 0 | **0.84** | 15 | 0 | 2 | 0 | 15 | **1.00** | 1 |
| CME and ADBE | | | | | | | | | | |
| Default | **0.04** | 0 | **0.01** | 15 | 0 | 2 | **0.07** | 15 | **0.00** | 1 |
| RRP unload | **10.29** | 0 | **0.87** | 15 | 0 | 2 | **0.01** | 15 | **3.50** | 1 |
| RP unload | **4.03** | 0 | **1.06** | 15 | **0.58** | 2 | **0.87** | 15 | **0.36** | 1 |

# 6   Case Study 2: pHluorin Based Assay

This case study is based on current pHluorin experimental work from the Cousin laboratory. The assay demonstrates repeated stimulus and recovery cycles, along

**Table 3.** Protocol events for the pHluorin based assay.

| Event | Stimulus | Start time (s) | Duration (s) |
|---|---|---|---|
| Rest | None | 0 | 1300 |
| Stimulus train 1 | Electrical (10 Hz) | 60 | 30 |
| Stimulus train 2 | Electrical (10 Hz) | 350 | 30 |
| Stimulus train 3 | Electrical (10 Hz) | 650 | 30 |
| Stimulus train 4 | Electrical (10 Hz) | 950 | 30 |
| Ammonium pulse | Chemical (NH$_4$Cl, 50 mM) | 1180 | 120 |

with the use of pHluorin specific stimulus, shown in Table 3. Again, real experimental observations have been used.

Final results of inference as shown in Fig. 5, gave the parameters shown in Table 4, with a distance measure of −7.14.

**Table 4.** Inference Results for pHluorin Assay. Values shown are rate constants $C$ $(s^{-1})$ and time delay constants $T$ (s) for each of the discrete state change rules, with inferred values shown in bold.

| Event | Exocytosis | | CME | | ADBE | | Budding | | Pool change | |
|---|---|---|---|---|---|---|---|---|---|---|
| | $C$ | $T$ | $C$ | $T$ | $C$ | $T$ | $C$ | $T$ | $C$ | $T$ |
| Rest | 0.01 | 0 | **2.59** | 15 | 0 | 2 | 0 | 15 | **10.34** | 0 |
| Stimulus train 1 | **4.81** | 0 | **2.59** | 15 | 0 | 2 | 0 | 15 | **10.34** | 0 |
| Stimulus train 2 | **8.12** | 0 | **2.59** | 15 | 0 | 2 | 0 | 15 | **10.34** | 0 |
| Stimulus train 3 | **3.68** | 0 | **2.59** | 15 | 0 | 2 | 0 | 15 | **10.34** | 0 |
| Stimulus train 4 | **5.59** | 0 | **2.59** | 15 | 0 | 2 | 0 | 15 | **10.34** | 0 |
| Ammonium pulse | 0.01 | 0 | **2.59** | 15 | 0 | 2 | 0 | 15 | **10.34** | 0 |

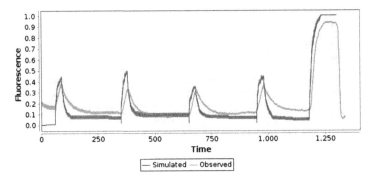

**Fig. 5.** Simulated fluorescence trace using the inferred parameters shown in Table 4, plotted against the real observed data used for parameter inference.

## 7    Conclusions

The FM-Sim application and its computational model of synaptic vesicle recycling has been demonstrated to work for a number of experimental assays. It provides the following benefits to both experimental and theoretical neuroscientists

- A user-friendly means of cataloguing experimental assays.
- An aid to experimental design by simulating the predicted effects of chemical modifiers on neurons prior to *in vitro* work.

– A means of gaining further insight into the kinetics of vesicle recycling by having numerous sets of observed data compared against a single kinetic model. This allows validation of the model and its rate parameters under a range of environmental conditions.

FM-Sim is currently being applied to a range of experimental assays within the Cousin laboratory at the University of Edinburgh. The research focus of this laboratory includes investigation of the mechanisms involved in synaptic vesicle recycling. Future work includes extending this library of experimental protocols, with supporting experimental data. In addition, we intend to cross-validate inferred experimental rates and models between compatible experiments, where environmental factors allow.

Further refinement of the model is also intended. In particular, ADBE derived vesicles are theorised to be processed by a number of sorting endosomes, modifying the complement of membrane proteins in the processed vesicle [6]. The computational model attempts to replicate these mechanisms to determine if such refinements improve the fit to observed data.

FM-Sim is available at http://homepages.inf.ed.ac.uk/s9269200/software/.

**Acknowledgements.** Thanks to the members of the Cousin group, in particular Sarah Gordon, for helpful discussions, and provision of experimental data. Thanks also to the reviewers of this paper for their helpful comments and suggestions.

This work was supported in part by grants EP/F500385/1 and BB/F529254/1 for the University of Edinburgh School of Informatics Doctoral Training Centre in Neuroinformatics and Computational Neuroscience (www.anc.ac.uk/dtc) from the UK Engineering and Physical Sciences Research Council (EPSRC), UK Biotechnology and Biological Sciences Research Council (BBSRC), and the UK Medical Research Council (MRC). The work has made use of resources provided by the Edinburgh Compute and Data Facility (ECDF; www.ecdf.ed.ac.uk), which has support from the eDIKT initiative (www.edikt.org.uk).

Stephen Gilmore is supported by the BBSRC SysMIC grant, BB/I014713/1.

# References

1. Andrieu, C., Doucet, A., Holenstein, R.: Particle Markov chain Monte Carlo methods. J. Roy. Stat. Soc. Ser. B (Stat. Methodol.) **72**(3), 269–342 (2010)
2. Aravanis, A., Pyle, J., Tsien, R.: Single synaptic vesicles fusing transiently and successively without loss of identity. Nature **423**(6940), 643–647 (2003)
3. Atluri, P.P., Ryan, T.A.: The kinetics of synaptic vesicle reacidification at hippocampal nerve terminals. J. Neurosci. **26**(8), 2313–2320 (2006)
4. Barrio, M., Burrage, K., Leier, A., Tian, T.: Oscillatory regulation of Hes1: discrete stochastic delay modelling and simulation. PLoS Comput. Biol. **2**(9), e117 (2006)
5. Cheung, G., Jupp, O., Cousin, M.: Activity-dependent bulk endocytosis and clathrin-dependent endocytosis replenish specific synaptic vesicle pools in central nerve terminals. J. Neurosci. **30**(24), 8151–8161 (2010)
6. Cheung, G., Cousin, M.A.: Adaptor protein complexes 1 and 3 are essential for generation of synaptic vesicles from activity-dependent bulk endosome. J. Neurosci. **32**(17), 6014–6023 (2012)

7. Clayton, E., Cousin, M.: Differential labelling of bulk endocytosis in nerve terminals by FM dyes. Neurochem. Int. **53**(3), 51–55 (2008)
8. Cousin, M.: Use of FM1-43 and other derivatives to investigate neuronal function. Curr. Protoc. Neurosci. **43**(2.6), 2.6.1–2.6.12 (2008)
9. Cousin, M.: Activity-dependent bulk synaptic vesicle endocytosis - a fast, high capacity membrane retrieval mechanism. Mol. Neurobiol. **39**(3), 185–189 (2009)
10. Fernandez-Alfonso, T., Ryan, T.A.: A heterogeneous "resting" pool of synaptic vesicles that is dynamically interchanged across boutons in mammalian cns synapses. Brain Cell Biol. **36**(1), 87–100 (2008)
11. Gabriel, T., García-Pérez, E., Mahfooz, K., Goñi, J., Martínez-Turrillas, R., Pérez-Otaño, I., Lo, D., Wesseling, J.: A new kinetic framework for synaptic vesicle trafficking tested in synapsin knock-outs. J. Neurosci. **31**(32), 11563–11577 (2011)
12. Gillespie, D.: Exact stochastic simulation of coupled chemical reactions. J. Phys. Chem. **81**(25), 2340–2361 (1977)
13. Golightly, A., Wilkinson, D.: Bayesian parameter inference for stochastic biochemical network models using particle Markov chain Monte Carlo. Interface Focus **1**(6), 807–820 (2011)
14. Granseth, B., Lagnado, L.: The role of endocytosis in regulating the strength of hippocampal synapses. J. Physiol. **586**(24), 5969–5982 (2008)
15. Granseth, B., Odermatt, B., Royle, S.J., Lagnado, L.: Clathrin-mediated endocytosis is the dominant mechanism of vesicle retrieval at hippocampal synapses. Neuron **51**(6), 773–786 (2006)
16. McMahon, H., Boucrot, E.: Molecular mechanism and physiological functions of clathrin-mediated endocytosis. Nat. Rev. Mol. Cell Biol. **12**(8), 517–533 (2011)
17. Miesenböck, G., De Angelis, D.A., Rothman, J.E.: Visualizing secretion and synaptic transmission with pH-sensitive green fluorescent proteins. Nature **394**(6689), 192–195 (1998)
18. Royle, S., Granseth, B., Odermatt, B., Derevier, A., Lagnado, L.: Imaging pHluorin-based probes at hippocampal synapses. Meth. Mol. Biol. **457**, 293–303 (2008)
19. Sankaranarayanan, S., De Angelis, D., Rothman, J., Ryan, T.: The use of pHluorins for optical measurements of presynaptic activity. Biophys. J. **79**(4), 2199–2208 (2000)
20. Wu, Y., Yeh, F., Mao, F., Chapman, E.: Biophysical characterization of styryl dye-membrane interactions. Biophys. J. **97**(1), 101–109 (2009)

# Author Index

Printed in the United States
By Bookmasters